I0029563

Mary Catherine Bateson (1939-2021). Photo courtesy of Sevanne Kassarjian.

Contents

Mary Catherine Bateson: Compositions in Living Cybernetics
Frederick Steier (Guest Editor) and Jane Jorgenson (Co-Editor)

The Artist for this issue is Matt Lorusso. Full color art at www.chkjournal.com. Artist's work can be found at https://micklorusso.net/.

Cover Art
Lorusso, Mick. (2006). *Messaggi del Cuore*. Cosecha de Luz Series, Energy Patterns. Painting. Oil on canvas. 200 x 300 cm.

CYBERNETICS & HUMAN KNOWING
A Journal of Second-Order Cybernetics, Autopoiesis & Cyber-Semiotics
ISSN: 0907-0877

Cybernetics and Human Knowing is a quarterly international multi- and trans-disciplinary journal focusing on second-order cybernetics and cybersemiotic approaches.

The journal is devoted to the new understandings of the self-organizing processes of information in human knowing that have arisen through the cybernetics of cybernetics, or second order cybernetics its relation and relevance to other interdisciplinary approaches such as C.S. Peirce's semiotics. This new development within the area of knowledge-directed processes is a non- disciplinary approach. Through the concept of self-reference it explores: cognition, communication and languaging in all of its manifestations; our understanding of organization and information in human, artificial and natural systems; and our understanding of understanding within the natural and social sciences, humanities, information and library science, and in social practices like design, education, organization, teaching, therapy, art, management and politics. Because of the interdisciplinary character articles are written in such a way that people from other domains can understand them. Articles from practitioners will be accepted in a special section. All articles are peer-reviewed.

Subscription Information

Price: Individual £80; Institutional: £188+VAT (online); £233 (online & print). 50% discount on full set of back volumes. Payment by cheque in £UK (pay Imprint Academic) to PO Box 200, Exeter EX5 5HY, UK; Visa/Mastercard/Amex
email: sandra@imprint.co.uk

Editor: Jeanette Bopry, Instructional Sciences, Ret. jeanette.bopry@gmail.com

Managing Editor: Carlos Vidales, University of Guadalajara, Mexico. morocoi@yahoo.com

Associate Editor: Sara Cannizzaro, Research Fellow, Centre for Computing and Social Responsibility (CCSR), De Montfort University, Leicester, UK. sblissa@gmail.com

Joint Art and Website Editor: Claudia Jacques, Westchester Community College, SUNY & Bronx Community College, CUNY. claudiajacquesmc@gmail.com; cj@claudiajacques.com

Book Review Editor: Bill Seaman, Duke University. bill.seaman@duke.edu

Special Topics Editor: Dirk Baecker, Witten/Herdecke University, Germany. dirk.baecker@uni-wh.de

Columnist: Lou Kauffman, University of Illinois–Chicago. kauffman@uic.edu

ASC Liaison: Ben Sweeting, University of Brighton, Brighton, UK. R.B.Sweeting@brighton.ac.uk

C&HK is indexed/abstracted in *Cabell's Journal* and *PsycInfo*
Journal homepage: **www.chkjournal.com**
Full text: **www.ingenta.com/journals/browse/imp**

Cybernetics and Human Knowing. Vol. 28 (2021), nos. 3-4, pp. 5–9

Foreword: Mary Catherine Bateson
Compositions in Living Cybernetics

Jane Jorgenson[1] and Frederick Steier[2]

In 1988 Mary Catherine Bateson was interviewed by the American journalist, Bill Moyers, for his "World of Ideas" television series about how gender roles are changing in the contemporary workplace. When Moyers asked her how we might go about changing how we think about the nature of evolving norms in contexts ranging from everyday life to cultural worldviews, Mary Catherine offered what she referred to as "a sort of parable" which centers on the Iditarod Trail Sled Dog Race in Alaska, and the strategy employed by Susan Butcher, a four-time Iditarod winner. Mary Catherine noted that Butcher differed from other competitors by lavishing care and protection on her dogs. In contrast to the dog mushers who achieved short-term gains by pushing the dogs to their limits, essentially exploiting them, Butcher won by focusing on sustaining them over the long haul. This strategy would most likely not work in shorter races, but Butcher's recognition of the Iditarod's extended time frame and her adaptive response, enabled her to win.

This recounting of Susan Butcher's story exemplifies s a central principle in Mary Catherine's work which was her ability to discern similarities across disparate domains and to use them to foster new ways of seeing for ourselves and for others. The content of this story coupled with the storytelling form is emblematic of how Mary Catherine made use of metaphor and analogy to advance and disseminate cybernetic ideas to both academic audiences and a wider public. The richness of her metaphors combined with their openness to different readings allow for multiple, intertwined meanings that a reader or listener might construct. While Mary Catherine's writing was grounded in the conceptual foundations of cybernetics, she didn't necessarily use the lexicon of cybernetics. She once said that she had a habit of purging abstractions from her books in order to allow stories to convey the kernels of her ideas and thus invite the reader in (Brockman, 2021). Her use of stories as well as her strong sense of audience undoubtedly contributed to the enormous reach of her work and her ability to provoke new ways of thinking about crucial human and environmental problems. In many ways, Mary Catherine could be understood to be the public face—the eyes and ears—of cybernetics.

Often asked to define cybernetics, Mary Catherine (2001) once wrote that, "Cybernetics can be a way of looking that cuts across fields, linking art and science and allowing us to move from a single organism to an ecosystem, from a forest to a university or a corporation, to recognize the essential recurrent patterns before taking

1. Department of Communication, University of South Florida. Email: jjorgens@usf.edu
2. School of Leadership Studies, Fielding Graduate University. Email: fsteier@gmail.com

action" (Bateson, 2001, p. 87). In her way of thinking, an intuitive grasp of resemblances across different domains and different scales of organization is essential for action and a path to wisdom. Indeed, Mary Catherine's work can be understood as a praxis of second-order cybernetics.

Throughout her career, she demonstrated, on one hand, a commitment to science-based thinking (Brockman, 2021) but this was combined with an openness to experiential participation in the world. Furthermore, she maintained a highly reflexive awareness of herself as observer who was a participant in the relational system she was observing. Mary Catherine wrote that it was her mother, Margaret Mead, who taught her "the lifelong habit of participant observation that has made me a social scientist and writer" (Brockman, 2005). Her style of participation was deeply embodied, as reflected, for example in her interest in joint performances as they unfolded in different contexts, whether in the intimate interactions between mothers and infants or the struggles of adults to communicate across cultural and language differences. Her 1993 piece, "Joint Performances: Improvisation in a Persian Garden," builds on her experience as a participant observer of a religious ritual in rural Iran who is simultaneously a parent engaged in caregiving for her young daughter. Here she offers a picture of the reciprocal processes at play when people improvise, including in tradition-rich contexts. In her later work on adult learning, Mary Catherine argued that the essence of learning was playful participation. As she said in an in an interview titled "Living as an Improvisational Art" with American author and journalist Krista Tippet, "Right through the life cycle, human beings remain playful—and play is a very important part of learning—and experimental" (Tippett, 2015, rebroadcast December 30, 2020).

The idea that we act our way into an understanding of seemingly incomprehensible events by recombining familiar (precomposed) and novel elements in new ways and then discern a pattern after the fact provided the framework for Bateson's exploration of women's lives in *Composing a Life* (Bateson, 1989). Here, building on a musical metaphor, she conceives of composing as process that unfolds over time in response to ever-changing contingencies. While composing can take different forms, she found greatest resonance in jazz improvisation as a metaphor for understanding women's creative adaptations to accelerating social change:

> Jazz exemplifies artistic activity that is at once individual and communal, performance that is both repetitive and innovative, each participant sometimes providing background support and sometimes flying free (Bateson, 1989, pp. 2–3).

In a similar spirit, we invite the readers to see the articles in this special issue as creative compositions in which recurring themes from Mary Catherine's work such as interdependence, continuity/discontinuity, reflexivity, metaphor, engaged observation and cybernetics itself are transformed by different voices and instruments across different registers. As with any jazz performance, the audience is integral to the scene. We hope these pieces provoke connections and improvisations for you as our audience, and that you will continue the conversation with us.

Mary Catherine Bateson passed away on January 2nd, 2021 and this collection of essays honors her life and work.

In "You Should Read this: Experiencing Our Own Metaphor," Paul Pangaro reflects on how he was introduced to Mary Catherine's first book, connecting with *Our Own Metaphor* through encounters with family members of prominent cyberneticians. Through these conversations, Paul recasts the events of the conference on which *Our Own Metaphor* was based, with the aim of bringing others into the story. In showing the relational reach of a meeting, Paul's paper can be seen as a kind of parallel process as he brings his own learning relationships into a new understanding of the meeting.

In "Embracing Relational Tensions in Research: What Mary Catherine Bateson Teaches Us," Jane Jorgenson orients to Mary Catherine as an ethnographer who was continually striving for self-reflexive awareness in her research encounters as a way of fostering a deeper understanding of others. Bateson's writing about her fieldwork, emphasizing the collaborative nature of research and the importance of listening to research partners at multiple levels, offers insights as Jane reconsiders a challenging research conversation.

Thomas Hylland Eriksen connects with the ideas of both Mary Catherine and her father, Gregory Bateson, as fellow anthropologists in "Anthropology in the Shadow of Anthropocene Overheating: Pantheistic Atheism and the Biosemiotic Turn." He argues that their interests in relationship, process, and the dynamics of living systems, deserve to brought more clearly into mainstream sociocultural anthropology. Thomas explores the affinities between Mary Catherine's perspective on the ecological embeddedness of human lives and biosemiotics, with the aim of developing a methodology for research on human life-worlds that attends adequately to the non-human world and offers new understandings of contemporary global challenges and crises of our time.

Alfonso Montuori's "Interdependence is the Key Issue: Mary Catherine and the Myth of Individualism" draws on examples from both interdisciplinary scholarship and popular culture to explore the idea of interdependence. Building on interdependence as a cornerstone of Mary Catherine Bateson's work, Alfonso illustrates how interdependence, and its conceptual partner of holistic thinking, is central in shaping a cybernetic understanding. Alfonso further develops the importance of interdependence as a path to processes of creativity and transformation.

Metalogues were central to both Gregory Bateson's *Steps to an Ecology of Mind* and also Mary Catherine and Gregory's *Angel's Fear*. Inspired by the metalogues and their emphasis on recursion, Hilary Keeney and Bradford Keeney's "Chasing the Mind and Body of Metalogue, Catching Recursive Frame Analysis" develops their forms and practices of Recursive Frame Analysis. Themes from the metalogues, including ideas of logical typing, emergence, and challenging assumptions, afford Hillary and Brad an opportunity to highlight the ways in which their Recursive Frame Analysis opens into processes of improvisation and possibilities of transformation, themes also central to Mary Catherine Bateson's work.

In "Cybernetics, Global Issues, and the Need for Systemic Wisdom: Mary Catherine Bateson and the Salzburg Global Seminar," William Reckmeyer builds on his collaboration with Mary Catherine Bateson in the Salzburg (Austria) Global Seminar in the years 2007-2010. He highlights the ways in which Mary Catherine brought cybernetic ideas to the international community in Salzburg as a way of fostering a conceptual basis for creating an informed and engaged global citizenry. Bill's recounting of Mary Catherine's work at Salzburg highlights her concern with ways of both understanding and then acting into complex systems, and the value in cybernetic approaches in doing so. In addition to the content of Mary Catherine's work, Bill stresses Mary Catherine's emphasis on building relationships for learning together.

Jude Lombardi and Larry Richards take the occasion of a letter sent by Mary Catherine Bateson to the American Society for Cybernetics on her being awarded the Norbert Wiener medal for lifetime achievement. The letter calls for a renewed appreciation of her father Gregory Bateson's talk, "From Versailles to Cybernetics," and argues strongly for the need for cyberneticians to bring their ecological principles to strengthen interdisciplinary dialogue while also transforming "our shared understandings into a new kind of common sense." In "Rembering a Message from Mary Catherine Bateson," Jude and Larry craft a compelling conversation around their ways of understanding the talk so that their conversational process illustrates the very principles Mary Catherine offers in the letter.

Frederick Steier's "Braiding Continuity and Improvisation: Let's Go Exploring," draws on Mary Catherine Bateson's recurring theme of the interplay of continuity and discontinuity in peoples' lives and in the lives of transdisciplinary ways of understanding, such as cybernetics. In developing a picture of that interplay, Fred highlights the importance of flexibility with a move toward an appreciation of improvisational exploration as a central theme of living cybernetics.

In addition to the essays, there are the regular featured columns and book review. In his column, Virtual Logic-Recursive Distinctions, Louis Kauffman introduces readers to recursive distinctioning. He offers a model and then discusses the cybernetics of describing *describing* in a context called *audioactivity*, linking to circularities for generating meaning for second-order cybernetics.

Goran Matic contributes this issue's American Society for Cybernetics column, "Acting Cybernetically in Complex Social Challenges: Designing for Sustainable Innovation." Matic advances perspectives on what it means to act cybernetically, inviting the reader to frame conversation as design strategy while considering design as ethics.

Devon Schiller offers this issue's book review, "The Face and the Machine," a review of Ksenia Fedorova's *Tactics of Interfacing: Encoding Affect in Art and Technology*.

This issue's featured artist is Mick Lorusso, M.F.A., a cross-disciplinary artist who, like M.C. Bateson, interweaves musings on molecules, cells, societies, and environments. His avant-guard and complex thinking coupled with his technological, artistic, scientific and research skills reflect an effective educator, in the classroom and through his artwork.

Lorusso has been a resident artist at interdisciplinary art programs, including PLAND / ISEA 2012: Machine Wilderness (Taos, NM, USA), the 2016 Rauschenberg Rising Waters Confab (Captiva, FL, USA), and Matza Aletsch 2017 in the Swiss Alps. At a residency in Schöppingen, Germany, Lorusso harnessed electricity-producing bacteria to illuminate a sculptural village, Microbial Schöppingen, which received a hybrid art honorary mention at Ars Electronica 2013. With training in microbiology and Kundalini Yoga and education in art at Colorado College (BA) and San Francisco Art Institute (MFA), he is a member of the UCLA Art|Sci Collective and has served as an instructor for the UCLA Sci|Art Nanolab Summer Institute since 2014. He is a graduate of the Waag Biohack Academy, and is a member of the international education platform Cosmic Labyrinth and the Mexican interdisciplinary cooperative XOCIARTEK. He is currently the STEAMWork Design Studio Coordinator at Westridge School in Pasadena, CA.

Guest editor for this issue is Frederick Steier, Professor in the School of Leadership Studies at Fielding Graduate University, and Emeritus, Dept. of Communication, University of South Florida. Co-editor is Jane Jorgenson, Professor of Communication at the University of South Florida

References

Bateson, M. C. (1990). *Composing a life*. New York: Penguin Books.

Bateson, M. C. (1993). Joint performance across cultures: Improvisation in a Persian Garden. *Text and Performance Quarterly*, 13(2), 113–121.

Bateson, M. C. (2001). The wisdom of recognition. *Cybernetics & Human Knowing, 8*(4), 87–90.

Brockman, J. (2005). *Curious minds: How a child becomes a scientist*. New York: Pantheon Books.

Brockman, J. (2021, January 18). Mary Catherine Bateson: Systems thinker. *Edge*. Retrieved December 31, 2021 from https://www.edge.org/conversation/mary-catherine-bateson

Tippett, K. (2015, October 1, rebroadcast December 31, 2020). Mary Catherine Bateson: Living as an Improvisational Art. *On Being with Krista Tippett* [podcast]. Retrieved December 31, 2021 from https://onbeing.org/programs/mary-catherine-bateson-living-as-an-improvisational-art/

Lorusso, Mick. (2008). *Dusk Mask Drawings* (detail). Urban Energy Drawing Series.

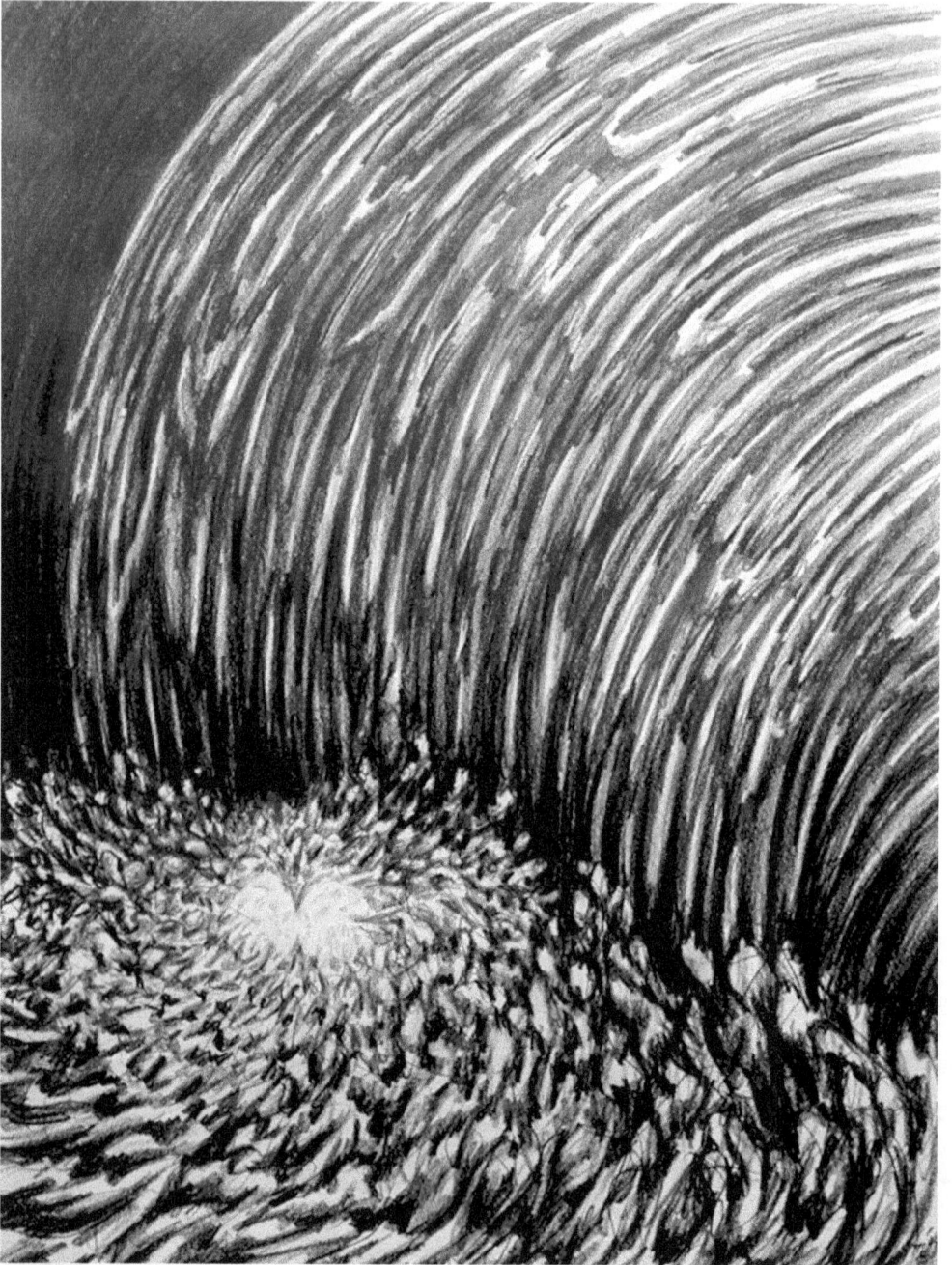

Lorusso, Mick. (2006-08). *Cascade,* Anima Mundi Series, Energy Patterns.
Drawing. Graphite on paper. 43 x 35 cm.

Cybernetics and Human Knowing. Vol. 28 (2021), nos. 3-4, pp. 11–21

You Should Read This
Experiencing *Our Own Metaphor*

Paul Pangaro[1]

This paper recalls the relationships I had with critical contributors but hidden figures in the history of cybernetics. They are pivotal to my understanding of cybernetics as well as my connection to other figures recognized as canonical to the field. My self-awareness of these decisive relationships began with reading *Our Own Metaphor* by Mary Catherine Bateson. The full significance of these relationships to my understanding of cybernetics grew through personal reflection, catalyzed by multiple encounters with multiple figures over many years. I write to give homage to Bateson and to all who brought not simply a parallel framing of cybernetics as inter-relational but the experience of the essence of our co-existence with others—and how this changed my relationship to cybernetics itself.

Keywords: Mary Catherine Bateson, second-order cybernetics, population crisis.

Intention

I write with inter-related intentions. One is to continue the self-reflection that has been so valuable to me and that began with Mary Catherine Bateson's writing.

Another intention is to introduce others to her writing, to spotlight what I find irreplaceable, so that her contribution and that of others may be better recognized, spread, and celebrated.

I'm not a historian but a passing participant-observer in the community of the *practice of cybernetics*. But I sense that, perhaps because Bateson was the daughter of parents who were two of the most prominent practitioners in their fields, and more likely because she was female, that during the main decades of her work she far from received the credence she deserved. It's a new era but there are still many wrongs to redress. Let me place her in the position of main exemplar here, but there are others who are also not well-enough recognized. *Our Own Metaphor* was for me a portal to the worlds inhabited by three additional and extraordinary human beings, not credited as part of the practice of cybernetics yet irreplaceably part of it. This paper is a partial remembrance of their stories.

By analogy to Ruth Bader Ginsberg, known as RGB for her personal power and influence (a towering figure also framed by her era and gender), herein I will respectfully refer to Bateson as MCB.

1. Human-Computer Interaction Institute, Carnegie Mellon University. President, American Society for Cybernetics. Email: paulpangaro@pangaro.com

First Encounter

"You should read this," Elizabeth said, "it'll tell you the human side." It was around 1977 that she put in my hands a first edition of *Our Own Metaphor* (Bateson, 1991). The author's name, Mary Catherine Bateson, was unfamiliar to me. "The daughter of Margaret Mead and Gregory Bateson." Those names I knew a little. "Mary Catherine is writing about a conference from her perspective. It's very second-order. Gordon was there and Warren was there, too. It was the last conference before Warren died." Gordon is Gordon Pask, Warren is Warren McCulloch. The conference was organized by Gregory Bateson and took place in 1968.

Now, Elizabeth is Elizabeth Poole Pask, Gordon's spouse. She was right, of course, about Bateson's text being an embodiment of second-order cybernetics—where the observer knows that the observer is inside the system under scrutiny—and also, therefore, that the observer is responsible for their observations (von Foerster 1991).

In such matters Elizabeth, as for all the great spouses of great cyberneticians that I have been fortunate to know, was always right.

We were in the living-dining room of their home on Montague Road, Gordon and Elizabeth's Edwardian house on the ritzy side of Richmond Hill in Richmond-upon-Thames, Surrey, just west of greater London. There were spotlights embedded in the ceiling, a theatrical modernization that worked to great effect on this occasion as well as whenever Gordon was holding forth in the presence of guests, richly fed by Elizabeth, not always willingly. (I report this as a fact. In no way do I reduce her importance as otherwise described here.)

I had first appeared on the Pask's threshold not long before. Like other acolytes of Pask and cybernetics who came for ideas and nourishment, I was hungry for alternatives to the mainstream, artificially-intelligent views of digital computation—those highly limited, if readily available, fashionable, and (still today) spectacular failures at making human-machine interactions richer, more organic, more humane. (I felt that Pask had the right models of interaction and intelligence while AI had the wrong ones [Pangaro, 2017]).

That was why I repeated the pilgrimage to Richmond Hill over the course of many years and how I learned that Elizabeth was a major complementing force to the powerful scientist in the double-breasted suit and proper bow tie, Dr Andrew Gordon Speedie Pask. Conversations with Elizabeth—Lizzie, as Gordon called her—were the beginning of a parallel yet braided education on the complementarities of rational description and emotional explanation. (Let's go all the way: the digital and the analog.) MCB's book was the first catalyst for this education. (Thank you, Lizzie.)

First Reading

In my first reading of *Our Own Metaphor* I found it unlike any conference report I'd ever seen. (*Report* is hardly a proper term, yet I struggle to replace it.) It is a remarkable work. MCB's clear-eyed explanations of both the concepts and arguments of the conference—the fundamentals of command versus control; linear versus circular-causal feedback; goals for a system vs. goals in a system; the necessary role of levels and hierarchies in systems that learn (this litany is far from complete)—deft, clear retellings are woven into her unflinching explanations of the emotional dynamics among the participants, of which she was one. At various points she notes, for example, that Warren "glared," others reacted "tensely," "very excitedly," or "spoke in rage—or perhaps contempt." MCB was herself "full of misgivings" at a late point in the conference when Gordon was about to speak. So often in this direct way her text evokes the emotions of the interaction, dimensions not simply missing from standard conference proceedings but consciously excised, as if we could or should ignore that part of our selves, the human-animal hormonal system that complicates everything we are and do together. (For explication of how this is a folly to ignore, see von Foerster, 1973.)

Starting only with Elizabeth and MCB, do I overgeneralize to say the women of cybernetics have much to bring us? Not at all. (And there are more to follow, read on.) The standard history of a field is about the occasions and ideas, not the emotions and conflicts. It was the ideas that first drew me into cybernetics, I thought. But MCB helped me become aware that it was the interpersonal dynamics, the emotional flux that was equally captivating. And whose second-order sensibility makes a core contribution to the field of cybernetics itself. (Thank you, MCB.)

First Cybernetics

When I first met Gordon Pask at MIT I didn't know so much of cybernetics. (Well, yes, that's true.) So where had I come from? I had finished an undergrad degree in 1974 by bridging the fields of humanities and computer science, seeking refuge in the former as the latter revealed itself to be so constricted. While I knew my goal was to program machines, to satisfy a desire (a need, really) for better interaction (Pangaro, 2020), the approach of MIT's artificial intelligence community felt inadequate, even inept in the context of its own declared goals.

So I gravitated to Jerry Lettvin and one of his postdocs, writing computer simulations and dynamic visualizations of how nerve impulses behaved as they flew out of the back end of neurons into giant dendritic trees connecting to thousands of other neurons (Raymond & Pangaro, 1977). Our goal was to fathom how some type of information processing was taking place as some of the electrical pulse trains continued and some were stopped at the trees' branches. You see, the impulse trains that came out of each of the billions of neurons in the human brain didn't come out of every branched axon at the far end, where they connected to 10,000 further neurons (Raymond & Pangaro, 1974). Surely something important was going on. And no one was looking at this then. (Or now. The neuron is considered all important.) This was cybernetics at its origins—focusing on the nervous system as a machine (a recurring topic in *Our Own Metaphor*).

Of course I knew that Jerry had worked with Warren McCulloch (and also, of course, with Humberto Maturana). Their names were often spoken in the lab and not just by Jerry. But it was from Jerry that I heard the compelling application of first-order feedback in such a natural context—in how the nervous system completes the loop of action and consequence in the world through sensing and comparing to possible goals and back to action. (For a particularly accessible recounting with wonderful hand-drawings, see Maturana, 1985.) So while Jerry had been talking cybernetics without much using that word, I was steeping in the cybernetic framework of structural coupling of the embodied, responsive organism and its environment.

Second Cybernetics

Then one day I glanced into the office of Nicholas Negroponte at MIT's Architecture Machine Group and I was riveted by the materialization of a small but powerfully intense Edwardian gentlemen, in double-breasted jacket and bow tie, smoking fragrant tobacco in an odd metal pipe. This was my first encounter with Pask. He spoke in a language—his own, yet using common words such as *topics, conversation,* and *agreement*—with highly specific, meticulous, rigorous meanings that took time for me to make *coherent.* (That was another of his coded terms.)

After conversations with him and conversations with myself to understand his dense writings, I realized his conversation theory was much closer to the sensibility of human beings than AI was or could ever be, while covering the same ground. From his papers and his theory, I realized there was a powerful alternative to the AI point of view, which was the only game in town at MIT.[2] Delving into Pask's cybernetics and then the cybernetics of those to whom he introduced me—von Foerster, Beer, Maturana, Glanville, Pedretti, both as human beings and for their work—it grew clear that it was time to leave the MIT PhD program and its familiar surroundings, conceptually and geographically, for this adventurous and more promising realm.

But I confess it was as much for the qualities of Pask as a human being—his intellectual intensity, his compact and powerful presence, his ability to connect to an individual at his (her, my, their) own level and to create an experience that was individualized to an extreme—that drew me to his work. (And to the person.)

Elizabeth was extremely canny and understood this about me. She had experienced her own, very different but analogous attraction to G, as the family called him. (Their daughters, Amanda and Hermione, were constant, intelligent, spirited presences in the household.) MCB helped me express this to myself and Elizabeth did also—to reach second-order—by opening me up to the emotional ties that bind us to our selves and others.

So Elizabeth was as much an influence in my days on Richmond Hill as G, because she understood the primacy of relationships. G did also but his written theory

---

2. Joe Weizenbaum was also at MIT but was not mainstream as was Minsky and Papert, the priesthood of AI at MIT; to counter their views was to bring retribution (which also befell Terry Winograd). Weizenbaum was concerned with the culture of digital programmers for bending away from human values.

was expressed (at least in that era, the 1970s and 1980s) in ascetic, abstract, formal terms, and most strictly as diagrams and equations.[3] But MCB had as her goal to expose the underlying emotional flux, which resonated with me deeply. (And brought the ideas alive.)

The Conference

Elizabeth put MCB's book in my hands because it was about relationships and the human side of some of the greats of cybernetics, an exceptional group here in an exceptional setting. The Wenner-Gren Foundation funded a conference of Gregory Bateson's design, which he titled "The Effects of Conscious Purpose on Human Adaptation." He also chose the 14 participants (including himself) who gathered over seven days in a castle named Burg Wartenstein in Austria in 1968. All fourteen were on the same footing, including MCB herself, one of only two women.[4] Anthropologist that she was (and child of parents both from that field), she wrote about the conference as a participant-observer, her feelings braided into her descriptions of the situations (specifically: who argued with whom about what). I was particularly anxious to read about Warren, for whom this was his last conference. He died in the month that I arrived at MIT in September 1969, already a mysterious legend.

Warren

Elizabeth had told me that Warren was a "true genius" and she distinguished him definitively from Norbert Wiener, whom she called a "manufactured genius" (E. Pask 1977, pers. comm.). A decade before I had read Wiener's two-volume autobiography in my MIT undergrad dorm room, so I volunteered to Elizabeth an interpretation of what she meant: that Wiener's domineering and obsessive father had manufactured him to be a genius. "Precisely," she replied. (Elizabeth was always right in such matters.) Elizabeth conjured for me her first encounter with Warren on a street in London, in cape and hat with a large feather, towering unmistakably above the throng, confirming his admonishment when she had asked beforehand, how would she know him? "Oh, you'll know me!" he bellowed into the phone before ringing off. (And so she did.)

I also knew about Warren from Jerry and others in the lab that they all once shared. At the celebration of Jerry's 60th birthday, co-workers and friends offered many tributes to each of them, lengthy stories about their work together and their personal relationships. When it seemed that the event was wrapping up, Jerry interrupted—no one had thought to reserve a slot in the schedule for him to speak—

---

3. Humberto Maturana thought this was because Pask was uncomfortable in his own body. Conversation with the author at a conference of the American Society for Cybernetics in Santa Cruz, California, 2002.
4. The second female participant at the conference was Gertrude Hendrix, described by MCB in an appendix as a mathematician and educator who presented on unverbalized awareness. Her work on the topic had led her to be introduced to Margaret Mead (and then to MCB or vice-versa). See https://www.depauw.edu/news-media/latest-news/details/23259/ retrieved November 13, 2021.

but he stood up aggressively and insisted on speaking, primarily about Warren and Walter Pitts, his very close colleagues and friends over many years. McCulloch and Pitts were the originators of considering neural nets as computational machines and proved their equivalence to Turing machines and therefore to digital computers (McCulloch & Pitts, 1943). I can still see Jerry pacing back and forth in the front of the room with his usual focus and clear arc of intention in his storytelling, chain smoking cigarettes, recalling how, when Warren and Walter began their collaboration, Warren was the poet—full of metaphor and historical reach—while Walter was the logician—requiring mathematics and specifics at every turn. However, Jerry said almost wistfully, over time in their relationship they traded roles, with Warren becoming the logician and Walter more poetical. Jerry said he missed them both very much. (This was clear from his voice and his demeanor.)

I recognize only now, thanks to MCB (on the path started by Elizabeth), that the emotional flux comprising our relationships is not a liminal space from which the rational emerges—rather, the flux of relationships is where all is engendered and embraced.

Rook in Old Lyme

I have always welcomed anything that brought Warren into closer view. (One superb, essential source is McCulloch, 1974.) In the mid-1980s I had the opportunity to visit Warren's widow, Rook McCulloch, on the McCulloch farm in Old Lyme, Connecticut. Just as Elizabeth was extraordinary, Rook was extraordinary, though in different ways, naturally. Rook to me was always warm and unselfish and inevitably kind and completely grounded. She had adopted many children, sometimes already adults, as I recall. I stayed overnight there in Warren and Rook's house on a few occasions, including once with Gordon. The kitchen was always replete with food, serving many. The rooms of the farmhouse all spoke of Rook. After dinner we went into a study with book shelves where Rook played a recording of a violin sonata. She told stories about the farm, about her friends and her children and a grandson. After silence in the room for several moments, Rook said that, from her childhood, silence in the conversation meant that an angel was coming through the room.

And the spirit of Warren was present, unmistakable. Walking outside at night on the farm with Gordon it was suddenly stormy and very, very windy and we were almost blown over, Gordon with his pipe no longer working and his cape swirling up around his head. He volunteered that this was the spirit of Warren himself coming through to visit. (Before he said this, it felt that way to me, too.) Another angel very present. (Thank you, Rook.)

Mai in Pescadero

So frequent in all these times, with Rook and Elizabeth and especially G, was the mention of Heinz von Foerster. (G spoke often of Warren as a father figure to him.) I spoke with Heinz at cybernetics conferences in the 1980s but knew him better from the late 1990s when I moved near to where he had retired, a small town called Pescadero, south of San Francisco, near the coast. My visits there were many and rich and vivid still, including, perhaps especially because of Mai, his spouse.

Mai once said to me, in that precise clipped English that was a feature from her German roots, that "Heinz – has a mind – like a crystal." But it's equally true that Mai herself had a mind like a crystal; and, unlike Heinz, she knew the art of concision. Heinz would speak for minutes on end, stories of cybernetics and cyberneticians, over one of Mai's amazing dinners (yes, imposed upon to offer dinner, even after generations of students during their time at the University of Illinois Urbana-Champaign, and now with friends and former colleagues dropping by in Pescadero).[5] Then, after a brief pause and in a few short sentences, Mai would bring her own summary, adding her own crisp insights, in a few sentences. Silence would follow, an angel coming through. (Thank you, Mai.)

Second Reading

(Did I digress?) MCB in her extraordinary account in *Our Own Metaphor* manages to out second-order the second-order geniuses who were there. Participant-observer to her core, every page speaks of her personal relationship to the expression of the ideas of others while working to be responsible to their intention and emotions.

A recurring theme across MCB's writing and lectures was the ecological crisis she speaks of so often in *Our Own Metaphor*, including her father's growing concern about it and how he expressed this in his invitation to the conference participants in the form of "Memorandum," which she quotes in full (G. Bateson, 1972/2000, pp. 446–453; quoted in Bateson, 1991, pp. 13–17). Gregory's concept for the conference was whether human consciousness perhaps especially as it is shaped in modern Western culture "might contain systematic distortions of view, which, when implemented by modern technology, become destructive of the balances between individual man, human society and the ecosystem of the planet" (G. Bateson, 1972/2000, p. 446; quoted in Bateson, 1991, pp. 12–13). (Prescient. Alarming. I didn't pay enough attention on first reading.)

On revisiting *Our Own Metaphor* to prepare a keynote with a spotlight on MCB's mother Margaret Mead (Pangaro, 2021), I was thunderstruck (a phrase I owe to Mai) by another way in which cybernetics could help one of the modern pandemics of ecology and inequality, climate and pollution, automation and artificial intelligence.[6] I had lost that it was hiding here in her book in plain sight, a beautiful implication from Pask's conversation theory, beautifully explicated by MCB, as follows.

---

5. Mai was famous for her dinners and especially for her Sachertorte as a sweet finish, known to be superior to those of the world-famous Cafe Mozart in Vienna.

6. Casting these global wicked problems as pandemics followed inevitably from the COVID-19 contagion that took hold worldwide in March of 2020. This led to an initiative dubbed #NewMacy Conversations, from which has followed a set of themes and on-going activities. See the evolving documentation at https://tinyurl.com/newmacy-docs.

Near the end of the conference Gordon offers some of his core concepts as a path forward for addressing one of the pandemics of our current millennium: the human population explosion that is choking the planet, all other species, and ourselves. This concern was not new even in 1968. Within the cybernetic community, Heinz (at least) famously wrote about it in 1960 (von Foerster, Mora, & Amiot, 1960) and the topic was very present during the conference in 1968, over which time the global population grew from roughly 3 billion to 3.5 billion. (Today, fifty years later it stands close to 8 billion, with estimates that the planet can only sustain 9 to 10 billion. Terrifying. Stultifying.)

As MCB retells,[7] Gordon begins, "Now I don't think that you could ever say that an individual should not reproduce. It would seem to me an essential part of the notion of 'consciousness' that an individual is that which wants to reproduce or perpetuate itself" (quoted in Bateson, 1977, p. 307).

Gordon goes on to speak in terms that are quite mechanical—right then he even calls himself a philosophical mechanic—but we can feel his obsession with rigor and generality (true to him being also a theorist) as he explains his framing of human beings in two complementary ways. One, we are physical, biological, organic creatures with a single brain. These he dubs mechanical individuals or *M-Individuals*. Two, as the consequence of processes executed in that single brain, there are a multitude of perspectives, points-of-view, frames for characterizing concepts or experiences. These perspectives need not be consistent with each other—they are even likely to conflict—but they must be stable enough to be persistent—in Pask's terms, *coherent*—and therefore they form memories. These he called psychological-individuals or *P-Individuals*.[8]

Put another way, "I am multitudes" (Whitman, 1892). (Today we would say a multiverse or pluriverse of individuals that inhabit our m-selves. And no, I refuse to mention Facebook's Metaverse.)

Pask argues that the will or even the obsession of individuals to reproduce is a biological fact; yet he offers an alternative: "I believe that the first thing we must do is redefine what we mean by an individual, get away from this idea of individual as heads. We can't get away—I don't want to get away—from the idea [of] *I*" (Bateson, 1977, p. 309).

Could this be a viable way to lessen population growth? Could we apply these tenets of cybernetics and conversation theory to devise a plan to effectively spread this way of thinking about being and living, to find a path to shared self-preservation, instead of species and ecology and planet destruction? (Isn't there a movement in here somewhere? Anyone up for that?)

---

7. Total anthropologist-researcher that she is, in her preface MCB careful explains that she made her transcripts from the original audio recordings while taking complete responsibility for any errors. Would that we had these recordings available today.

8. My colleague Claudia L'Amoreaux has renamed these p-selves which I feel holds the same meaning while capturing more of the intention.

Change/Not-Change

Here is my strongest evidence of MCB's rare and invaluable artistry for explaining core concepts of cybernetics. She claims at one point (Bateson, 1977, p. 204) that all change can be understood as a desire to bring about constancy. (Without further explanation and the specificity of first-order systems, that reads as a contradiction or a koan.) She explains: A desire to bring about constancy is the desire to regulate, to hold something (relatively) constant, to *control*, here meaning to attempt to regulate some aspect of one's environment. *Constancy* here means within acceptable boundaries, within one's current goal. The means to this end, cybernetics says, is by changing our actions in a changing world or environment in order to adapt to new conditions—to conserve that which we wish to conserve.

Complementarily, all desire to bring about constancy can be seen as a desire for change.

This is such a beautiful, even breathtaking way of talking about first-order cybernetics—by popping out to an observer's view and by seeing the first-order dynamics in terms of change/not-change from a second-order frame. It holds the tension between change and not-change; it avoids letting the pair reduce to a trivial dichotomy. This is typical of the power by which MCB is able to recast ideas, in this case, about cybernetic goals. In other moments she recasts apparently rational exchanges into the anthropology of the conference with its human emotionality; revisits and lays bare many of the concepts developed by her mother and her father; and, from the stance of a practitioner as well as a powerful writer, she is able to contribute to our better understandings of relations that are both emotional and rational. (We as humans perform both. We are both.)

MCB embraces the individuals that embody the conference and makes them an example of community and society and human connection and synergies—all grounded in emotion. Hence—wait for it—the construction in her book's title, *Our Own Metaphor*. Toward the very end of the conference, Gregory speaks that phrase—articulating that the interactions of the participants enact the concepts of the conference, embody the emotional flux of the conversations. They—the P-Individuals and the conversations they are—become their own metaphor. Like her mother and her father, MCB was able to make a direct and profound contribution to the anthropology of cybernetics, that seems to me as great as any other. (Read that again, please.)

Being of Cybernetics

Elizabeth, Mai, and Rook came to cybernetics by relationship rather than profession. It must be credited that Elizabeth consistently toiled on the business side of the Pask's non-profit and profit companies, and she contributed to the commercial contracts as well as many other adventures with G (read her evocative, breezy, but weighty and frank prose in Pask, 1993). (She was also pivotal in a courtroom once, for bringing a favorable verdict in a dispute over employee taxes.)

But whatever room they were in and whatever role they seemed to hold by convention of the times, all these women were part of the practice of cybernetics because they lived with it, lived within it, and co-made it with their spouses and the extended community they nourished, literally and spiritually. Their contributions must not be lost and can be celebrated much more. And there are many others; for one, the prodigious work of Annetta Pedretti (Pedretti 1981). I hope the contributions of all will be more recognized, better spread, and greatly celebrated. That has been one of my intentions here. Another has been to express my life-long gratitude.

Thank you, Mai.
Thank you, Rook.
Thank you, Lizzie.

And MCB, Catherine, as I might address her, continues to bring us all together in emotion and with love. From a forty-plus-year perspective, I see that *Our Own Metaphor* is an exceptional work. Scope, depth, heart, soul. Clarity, compassion, love. Timeliness—just one example, population growth—perhaps the core of all the global challenges that have only and will only continue to become more acute. (It cannot be too strongly said: We must listen better.)

Thank you, Catherine.

Afterward

At a small Thanksgiving gathering in the days I am writing this, I am asked in two distinct and unconnected conversations with bright and curious 30 year-olds, what book might be a good introduction to cybernetics? (I can't recall how the question arose at those two moments but I took it as a hopeful sign.) *Our Own Metaphor* can be that introduction, I say to them and to everyone. "You should read it." And I think to myself, I wish that I could know your feelings from your own second reading, forty-years hence.

Acknowledgements

The author wishes to thank Jane Jorgenson for her suggestions and Fred Steier for his patience, encouragement, and comments during the writing of this article.

References

Bateson, G. (2000). *Steps to an ecology of mind* (introduction by M. C. Bateson). University of Chicago Press. (Originally published in 1972)
Bateson, M. C. (1991). *Our own metaphor.* Smithsonian Institution Press. (Originally published by Knopf in 1972)
Lettvin, J. Y., & Maturana, H. (1959). What the frog's eye tells the frog's brain. *Proceedings of the IRE* [Institute of Radio Engineers], *47*(11), 1940–1951.

Maturana, H. (1985). What is it to see? *Cybernetic, Publication of the American Society for Cybernetics, 1*(1), 59–76 [Reprint]. (Originally published in *Arch Biol Med Exp, 16,* 255–269 in 1983). Retrieved November 13, 2021 from https://drive.google.com/file/d/1ur6fFuM4H2uZmm5CTDFYqfCkSHXIVxVe/view?usp=sharing

McCulloch, W. (1974). Recollections of the many sources of cybernetics. *ASC Forum 6*(2), 5–16.

McCulloch, W., & Pitts, W. (1988). A logical calculus of the ideas immanent in nervous activity. In embodiments of mind (pp. 19–39) [Reprint]. Cambridge, MA: The MIT Press. (Originally published in the *Bulletin of Mathematical Physics, 5,* 115–133 in 1943)

Pangaro, P. (2017). Questions for conversation theory or conversation theory in one hour. *Kybernetes, 46*(9), 1578–1587.

Pangaro, P., (2020). Winky Dink and me. In J. Chapman (Ed.), For the love of cybernetics: Personal narratives by cyberneticians (pp. 38–48). Routledge.

Pangaro, P. (2021). *Margaret's elegant question: #NewMacy in the 21st century.* Keynote given at WOSC 2021(World Organization for Systems and Cybernetics), September 27-29, 2021 in Moscow [delivered virtually September 29, 2021]. Retrieved November 13, 2021 from https://www.pangaro.com/wosc2021keynote/

Pask, E., (1993). Today has been going on for a very long time. In *Gordon Pask, A Festschrift* [Special issue]. *Systems Research Journal, 10*(3), (pp. 143–147).

Pedretti, A. (1981). *The cybernetics of language.* PhD Thesis, Brunel University. Retrieved November 13, 2021 from http://bura.brunel.ac.uk/handle/2438/5311

Raymond, S., & Pangaro, P. (1974). Nerve threshold and intermittent condition [16-minute computer-generated film]. Retrieved November 13, 2021 from https://vimeo.com/31219533

Raymond, S., & Pangaro, P. (1977). Mediation of impulse conduction in axons by threshold changes. *Society for Neurosciences, II,* 417.

von Foerster, H. (2003). On constructing a reality. Reprinted in *Understanding understanding* (pp. 211–227). Springer. (Originally published in F. E. Preiser [Ed.], *Environmental design research, vol. 2* [pp. 35–46] by Dowden, Hutchison & Ross, 1973)

von Foerster, H. (2003). Ethics and second-order cybernetics. In *Understanding understanding* (pp. 287–295). Springer. (Originally published in French in 1991)

von Foerster, H., Mora, P. M., & Amiot, L. W. (1960). Doomsday: Friday, 13 November, A.D. 2026. *Science, 132*(3436), 1291–1295.

Whitman, W. (1892). Song of myself. Retrieved November 13, 2021 from https://www.poetryfoundation.org/poems/45477/song-of-myself-1892-version

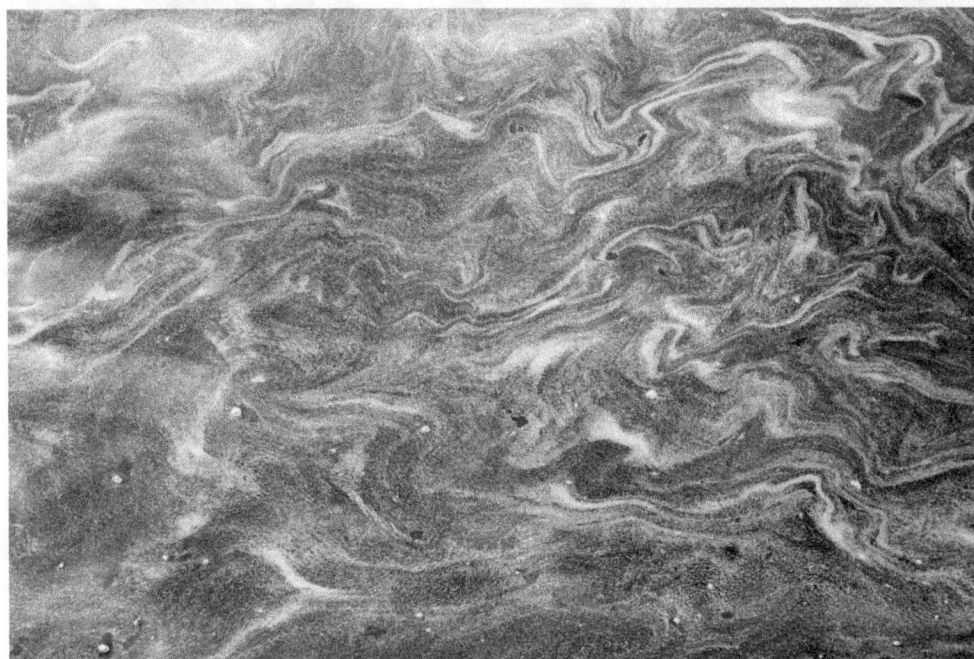

Lorusso, Mick. (2014). *Algae Hair.* Vodnik's Cells Series. Ecological Interactions. Microbial.
Photograph. 22 x 30 cm.

Lorusso, Mick. (2006-08). *Emanating Field*. Anima Mundi Series, Energy Patterns.
Drawing. Graphite on paper. 43 x 35 cm.

Cybernetics and Human Knowing. Vol. 28 (2021), nos. 3-4, pp. 23–31

Embracing Relational Tensions in Research
What Mary Catherine Bateson Teaches Us

Jane Jorgenson[1]

Mary Catherine Bateson's accounts of her fieldwork experiences afford rich opportunities for reflecting on the nature of research processes as interpersonal encounters. In this essay I consider several examples from different periods of Bateson's ethnographic writing that exemplify her reflexive awareness. Her work provides an ethical grounding for understanding a recent research experience and it offers insights for fostering a relational research practice by creating encounters characterized by deeper dialogic engagement with the participants.

Keywords: Mary Catherine Bateson; research relationships, reflexivity, work-family conflict; narrative interviews

A connecting thread throughout Mary Catherine Bateson's ethnographic writing is her sensitivity to research relationships and her awareness of her role as a researcher in setting the terms of the fieldwork situations she is trying to understand. Her anthropological work draws on experiences in a variety of international contexts during different periods in her life. Across these different projects she faced a dilemma that is common in fieldwork concerning the researcher's role in the research process. What researchers see and hear in the field is influenced by who they are—their identities, backgrounds and prior assumptions all affect what they notice and interpret as significant. Further, how the research participants view the researcher and her objectives, whether as a friend, a well-meaning outsider or a detached expert, complicates the interpretation of data and eventual claims that are made. The observed are also observing. The dilemma for many researchers is how to adequately acknowledge the influence of these observer effects, that is, how to incorporate the researcher's personal standpoints as well as the participants' views of the researcher in ways that enrich understanding without reinforcing the researcher's authority at the expense of participant voices.

In many ways her work anticipated the shift within anthropology in the following decades to a more self-critical stance as the risks of essentializing cultural groups and the unequal authority of the researcher and the "subjects" became central preoccupations (Clifford & Marcus, 1986; Marcus & Fischer, 1986) Rather than treating cultural differences as intrinsic, Bateson took a second-order view, reminding us that cultural differences are observers' constructions, expressions of comparisons based on the application of a priori categories. As early as the 1960s, her writing included moments of uncomfortable self-analysis, for example when she acknowledged feeling tensions as an American newcomer researching barrio communities in the Philippines. Yet through her analysis, the tensions become

---

1. Department of Communication, University of South Florida. Email: jjorgens@usf.edu

productive, providing a gateway to understanding rather than being a hurdle to overcome.

Reading Bateson's work affords rich opportunities for reflecting on how research is at its basis an interpersonal process marked by recursive perspective taking relations. She offers us unique insights into how researchers construct invitational frames in order to learn from others; how those others, the participants, respond to those invitations; and how the researcher's eventual findings are contingent on the emerging frames. These questions have been important to me in my current project interviewing women in different occupations about how they balance the commitments of paid work with caregiving for family members. Bateson's writing, by making explicit the relational processes through which her descriptions and analyses of culture are derived, has heightened my awareness of how the knowledge we arrive at is shaped by the dynamic, contingent process of the research encounter.

I begin this essay by highlighting several examples from different periods of Bateson's scholarship. I think of these as trail markers, each exemplifying in a distinct way her awareness of the research context and of herself as a participating member. For me, Bateson's work illuminates, not by explaining or prescribing a particular method she has used, but by pointing out opportunities for meta-reflection. Her writing has given me resources to think with as I try to make sense of women's stories of the trade-offs they have made in their working lives and, what is especially problematic, to understand the potential for my questions to undermine their established ways of coping. In the second part of this essay I reflect on an encounter with one woman, a research participant, whose depth of emotional engagement in the research conversation I was not fully prepared for. Bateson's work provides an ethical grounding as I consider the unanticipated effects my questions had on the interaction that emerged. In the final section of the essay I consider some of the implications of Bateson's work for how we might engage differently with others in research to achieve more multi-vocal perspectives.

Marking the Trail: Three Examples

Bateson often describes her experience of encountering cultural differences in narrative terms. Rather than presenting her observations as straightforward empirical "truths," she situates them in the context of her fieldwork relationships and social interactions, and includes her evolving emotional reactions as a central part of the story. In an early article based on her fieldwork experiences in the Philippines, she writes about the circumstances in which she learned about bereavement and its expressions in Filipino culture (Bateson, 1966). On one level this piece is about differences in how grieving and consoling are enacted in American and Filipino contexts. However her account also offers a second-order perspective by dwelling on her awareness of her position as an American newcomer and how this vantage point frames her observations.

The article is structured around a series of three fieldwork experiences. In the first, Bateson listens to a conversation between two women about the recent death of the second woman's son, and she notes her own outraged reaction to the seeming insensitivity of the friend who keeps probing the weeping mother for details about the death. Bateson takes her own feelings of anger as a cue for further inquiry into the nature of tact as a cultural construct. In a second episode she describes the experience of attending a vigil following the death of a neighbor in the barrio. Here she observes her feelings of embarrassment over entering the bereaved family's home, and how they seem at odds with the convivial spirit of the gathering. Finally a third more personal moment touches on the loss of her baby, born prematurely in Manila. Here, Bateson reflects on how her recent learning about the cultural dimensions of grieving enabled her to gratefully receive the consolations offered by Filipino friends rather than being alienated by them. Bateson's personal reactions and struggles during fieldwork are a key part of all these stories because she uses them as paths to ethnographic insight. As she notes, they complicate what might otherwise turn into totalizing descriptions of a culture or overly simplified bicultural comparisons. Part of her reflexive approach is her continuing awareness of difference between herself and those she is learning from. Throughout this account she acknowledges her multiple vantage points as a newcomer (who is visibly different), unfamiliar with the local practices and their meanings, and also as someone who is newly vulnerable through her recent loss. Through her exploration of these multiple identities, she conveys to the reader her evolving experience of otherness.

A second example of her reflexive awareness is found in Bateson's story of her arrival in Iran with her husband and two year old daughter in 1972 (Bateson, 1993). Here she recounts accepting an invitation to attend an Islamic observance, a ritual of sacrifice, in the countryside outside Tehran. Bringing her daughter, Vanni, along on the outing, she again finds herself negotiating multiple roles; as an anthropologist she feels drawn into an engagement with the details of the ritual (the words and gestures and the symbolism of the garden setting), while as a parent, she is narrating the unfolding events to try to make them intelligible to her young daughter. Much of this piece centers on the tension she feels as she moves between various roles: participating in the scene as a mother and a guest while observing as an anthropologist. Her solution to the tensions is to embrace the role of performing participant in the occasion and out of this she develops the idea of joint improvisation as a model of intercultural contact. In the absence of knowing the rules of the occasion or sharing a common code, she is forced to improvise, to participate in concert with the others who are present, always in a spontaneous way but with an attitude of good will.

My third example comes from the book, *Composing a Life* (1990), which takes the form of a memoir rather than a research account, in which Bateson reflects on her own life along with the lives of four women she admires. The book is based on her in-depth interviews with friends who she regards not only as women of achievement, but as exemplars of resilience who have used the disruptions and setbacks in their lives as the impetus for new learning. As she describes the process of gathering their stories, we see how Bateson sought to create as much as possible a nonhierarchical encounter through a series of extended conversations with each narrator, an approach designed to created space for each one to follow her own sense of direction.

It is clear that Bateson orients to this material in least two ways. First, she uses it as an entry point for understanding each woman's life world and the ways each individual has grappled with gender-based assumptions and inequities. Secondly, she is sensitive to the conversations themselves as storytelling occasions. She notes early in the book that in the process of verbalizing their experience the narrators are reflecting on and making sense of events to produce what they hope will be a culturally intelligible plot line. The retrospective nature of these accounts is key, for in responding to her invitation to tell their stories, the narrators are often editing and reinterpreting the past so that the events narrated are not necessarily as real or objective as the interview form suggests. One implication of this is that the work of crafting of a coherent life story can be a generative process by surfacing possibilities for creative adaptation and growth.

A Research Example: Narrative Struggles of a Working Mother

For the last several years I have been working with my collaborator, Annis Golden, to understand the impacts of new technologies in accelerating and intensifying working life. The specific focus of our project was on how increasing demands for employees' round-the-clock availability and reachability for work may affect working mothers' ability to manage their caregiving commitments. For this project we located a small number of women who were willing to share with us in detail their day-to-day experiences as they were lived, felt and interpreted. These women were diverse in their occupations but all were balancing paid work with caregiving responsibilities for children, in some cases also caring for parents or other family members. In conversational interviews we asked a series of questions on topics ranging from career history, household living arrangements, description of a typical work day, and in what ways they regarded work and family to be separate and segmented versus integrated, overlapping worlds. Our questions were open-ended, designed to allow the narrators freedom to define the scope of their responses so that the conversation can flow while also ensuring that common themes would be addressed across all the interviews. We did not consider these questions to be particularly intrusive or emotionally charged.

The example I focus on here is from my interview with Patricia, a 32 year-old college administrator who is married and the mother of with a 5-year old son. Since Patricia and I worked at the same university, we arranged to meet on a weekday in a conference room in my department for the interview. Like many women, Patricia found it hard to combine work and motherhood in a way that was satisfying to her, but the opportunity to talk about her experiences with someone and perhaps create benefit for others in the process seemed to have motivated her to participate. The notes I jotted immediately afterwards reflected my feeling that it had been a very successful interview in that Patricia had talked candidly and at length.

A key aspect of any interview context is its sequential organization. I began by outlining the purpose of the project: to shed light on the underlying sources of work-family conflict by exploring working mothers' perspectives, with the hope that our findings would, eventually, contribute to improved policies and programs. I also reviewed with her the Human Subjects consent process and we joked about the legalistic language of the consent form. Then I turned to the topic at hand, starting gradually as interviewers tend to do, by asking easier-to-answer questions to establish a sense of rapport before raising more complex or emotionally laden topics. I started by asking Patricia to describe her career path into higher education. She moved easily into a narrative about how her interests in educational leadership were formed in college. After college graduation and several years working in the insurance field, these interests were reenergized when she began volunteering on week-ends as an advisor to a college sorority. In the course of this volunteer work, she had an epiphany: that even though she was spending 12-hour days to drive to the volunteer work and back home, she found herself looking forward to her Sundays more than the rest of the week. "That's the moment where I realized, 'wow, maybe my calling is student affairs.'"

In reading the transcript of this exchange afterwards, I was struck by Patricia's idealism, her conviction that her work in student affairs connected her to a higher purpose. She presented her career selection as a deliberate choice grounded in her deepest meanings and expressive of her unique talents. Similar to the way many middle class Americans talk about their work (Ciulla, 2000), Patricia felt her work was more than just a job but a calling. In phrasing my question I had drawn on the familiar metaphor of career as a path, which tends to evoke the image of career choice as a process of continuous goal-directed upward movement, what Bateson (1992) has called a *steady achievement narrative*. But Patricia's story is a bit different by emphasizing a chance experience (a volunteer job) that led to a moment of self-discovery. Research on career storytelling has shown that women often present their careers as being shaped by accident and coincidence in contrast to many men's stories organized around the pursuit of clear and ambitious goals (Wagner & Wodak, 2006).

In the next portion of the interview I asked Patricia to describe a typical day with the idea of learning about the variety of roles she filled on a daily basis and the limitations under which she operated. At this point, she seemed to express more ambivalence, not about her career field but about the job itself. She walked me through the details of long work days running between meetings, often working on week-ends, conveying a sense of overwhelming demands on her time, which she summed up by saying that she doesn't cook (her husband does most of the cooking) and the family "eats out a lot which is terrible, but that's how you manage life sometimes." The final comment about eating out was only a brief aside but it reminded me of the strong cultural connection between good mothering and meal provision, and the predicament this creates (see e.g., Kinser, 2017). Although I didn't comment, I was reminded of all the effort involved in attempting to provide nutritious home-cooked dinners when my children were young.

To develop a picture of the place of work in relation to her other life goals, I then posed some additional open-ended questions: What does work mean to you? (Is it mainly a paycheck?); if you weren't in your current job is there anything would you miss? Would you prefer not to work at all if that was an option? These were followed by questions about how she made use of various technologies to work from home and to stay in touch with family while at work. But it was the initial question about work meanings that prompted Patricia to launch into a long narrative that began with her acknowledging her internal conflict: "In all honesty, if you would've asked me this four, maybe even three years ago, I would've told you it was my passion. I would've told you it was my purpose and the reason I got up every morning. Right now I'm not feeling that anymore." She went on to describe how she spent several years investing herself in her job with the understanding that she was being groomed for a more senior position. Then when the time came, the promised promotion was denied. Whereas her earlier story was about discovering what makes work meaningful (through a temporary volunteer experience), this episode centered on how she felt betrayed by the organization. She said the lesson from this experience was that "the people you think are in your corner are not." Patricia continued to talk about the insecurities she felt in relation to both motherhood and her career. For her, the impact of the setback was to foreground her inability to negotiate the right balance between family commitments and paid work. As she said,

> I essentially missed a lot of time … until my son was about three, I missed a lot of his life because I dedicated so much to work. So, now work is [only] a paycheck and it's like what do I need to do to be able to have a good evaluation so I can keep my job.

In spite of the deep connection to her profession expressed in her earlier response, it was obvious that this question provoked distress. Because her day-to-day life was so densely filled with work obligations, Patricia had to ration time for her son, for example by saving vacation days to use when he was too sick to go to school. She became openly emotional as she expressed her regret for having missed significant moments in his life. How, she wondered tearfully, would it be when he entered school in a few months; would she would be able to get home in time to help him with his homework? "How do I make sure that I'm active in his life in ways that I haven't been able to be over the last four years?"

Organization theorist Karl Weick's sensemaking theory offers a lens for understanding Patricia's struggles. He writes that the disruption of a planned future, including a career trajectory, is often an impetus for sensemaking (Weick, 1995). At such moments when identities are called into question (for example, the identity of *valued employee*), a person may try to restory events. Facing an obstacle to advancement she might find ways to rationalize and thus maintain a sense of self worth, saying, in essence, "that job I was turned down for wouldn't have been right for me anyway." But career disruption is harder to rationalize in situations where work and family spheres are so deeply enmeshed. The painful bind for Patricia lies in seeing her career choices as having directly impeded her ability to meet her expectations of good mothering. She expressed the dilemma this way:

> If I could, I would totally stay home [rather than work], but I've always been a very, very ambitious person, very—so, I'm having this really hard—I almost feel like I'm having an identity crisis. I'm not who I've always been and I'm okay with that, but I'm not okay with it at the same.

Understanding herself as an ambitious person, Patricia has difficulty naming the confusion she feels, eventually labeling it an identity crisis. "I am not who I have always been and it's okay but not okay." As Bateson found in her conversations with her friends, individuals can sometimes respond to setbacks by reframing events creatively so as to adapt to new circumstances. The nonlinear career stories as featured in *Composing a Life* are important to draw attention to precisely because they fall outside the culture's dominant narrative of success. But Patricia's account reveals that positive reframing is difficult without narrative resources, or stories, to provide frameworks for diagnosing the problem and guiding future action.

Rebalancing Interpretive Authority in Research Relationships

Inviting women to tell their stories is an opportunity to pay tribute to their subjectivity and creativity. By listening to their first-person accounts, we have the possibility of revealing socially important standpoints which have been overlooked and of revising received knowledge about their lives (Gluck & Patai, 1991). The narrators themselves can benefit from the storytelling process if they feel it validates the importance of their lives or normalizes an experience they had assumed was unique; in sharing their stories, they can arrive at new self-understandings (Opsal et al., 2016). Yet now months later, I am not sure what Patricia took away from this experience. From a research point of view, her account has many hallmarks of what Wagner and Wodak (2006) call a *good interview* with moments of remembering, reflecting, celebrating and regretting. Although she became openly emotional during the process of sharing experiences, especially those that didn't fit with dominant narratives of the good mother or the enterprising employee, she seemed anxious to tell those parts of the story. In response to her sharing, I mostly just listened, occasionally interjecting a word to convey the connection I felt to her feelings of loss and frustration. As we were ending and saying goodbye, she told me she enjoyed our conversation but I know that some of my questions prompted feelings of sadness and grief.

As feminist oral historians Sherna Berger Gluck and Daphne Patai (1991) have argued, the promise of collecting women's stories lies in bringing to light the social conditions and institutional constraints that inhibit positive change in their lives. They and other scholars are grappling with the ethical challenges of the research process, warning of the possibility for research to become manipulative when the conversation creates a false sense of intimacy that gets beyond a narrator's defenses (Brinkmann & Kvale, 2005). As I found out, questions may land differently than we intend, as they did in Patricia's case by surfacing profound doubts and regrets about her own actions. Christina Sinding and Jane Aronson (2003) suggest that a risk of any narrative inquiry is that the storytelling will unsettle accommodations, in other words, expose the tacit strategies a narrator may have developed in coping with difficult circumstances. Our questions can unwittingly encourage a participant to put words to uncertainties, and by doing so threaten the anchorage points of their identity.

Aronson and Sinding offer a way to repair the disruptions created by our questions, suggesting that we think of the interview as a reciprocal dialogue to which the informant contributes from a place of expertise so that the conversation becomes an opportunity for mutual consciousness-raising. The philosopher Linda Alcoff also considers dialogue the most promising available framework for addressing problems of representing the subjectivities of others. Ethical engagement, she writes, involves speaking *with* rather than speaking *to* or *for* others (Alcoff, 1992) and Bateson would surely agree with this. Her accounts of her fieldwork remind us of the intensely collaborative nature of any research task and the importance of offering invitational frames for others, even if others reject the frame or infer a different meaning from what we intend. This is exemplified in her notion of joint improvisation, which is distinguished by a stance of deep receptivity to others combined with a willingness to respond in the moment and always with good will. In this spirit, I wonder how it would have been to continue the dialogue with Patricia. Perhaps we could have continued the discussion of her stories in further conversations. I could have sought feedback by sharing my interpretations with her and asking her for her interpretations of my interpretations. As one small example, I might try to get her perspective on something I noticed in her story which was the stress created by an institutional policy that requires her to save up vacation days to use when her son is sick. Together we could explore our premises behind our interpretations and those conversations could then become part of the final product. By extending the dialogue and folding it into the circle of interpretation it may be possible to create a process that is less unsettling and of greater use to participants, as well as enabling us, the researchers, to reach a more adequately contextualized understanding of their lives.

An important part of Bateson's legacy is her commitment to human connection; as she reminds us, "the encounter with persons, one by one, rather than categories and generalities, is still the best way to cross lines of strangeness" (Bateson, 2000, p. 81). It seems likely that she and the friends who participated in *Composing a Life* would have continued to talk about their stories over time and in the context of the book's later reception. Taken together her work invites us to consider how we might create research encounters characterized by deeper dialogic engagement with participants and that honors the communicative and emotional work they invest in research conversations.

References

Alcoff, L. (1992). The problem of speaking for others. *Cultural Critique, 20*, 5–32.
Bateson, M.C. (1966). Insight in a bicultural context. *Philippine Studies, 16*(4), 605–621.

Bateson, M.C. (1990). *Composing a life*. New York: Penguin Books.

Bateson, M.C. (1992). The construction of continuity. In S. Srivastva & R. Fry (Eds.), *Executive and organizational continuity* (pp. 27–39). San Francisco, CA: Jossey-Bass.

Bateson, M.C. (1993). Joint performance across cultures: Improvisation in a Persian Garden. *Text and Performance Quarterly, 13*(2), 113–121.

Bateson, M.C. (2000). *Full circles, overlapping lives: Culture and generation in transition*. New York: Random House.

Brinkmann, S. & Kvale, S. (2005) Confronting the ethics of qualitative research. *Journal of Constructivist Psychology 18*(2), 157–181.

Ciulla, J. (2000). *The working life: The promise and betrayal of modern work*. New York: Three Rivers Press.

Clifford, J. & Marcus, G. (1986). *Writing culture: The poetics and politics of ethnography*. Berkeley, CA: University of California Press.

Gluck, S. B. & Patai, D. (1991). *Women's words: The feminist practice of oral history*. London: Routledge.

Kinser, A. (2017). Fixing food to fix families: Feeding risk discourse and the family meal. *Women's Studies in Communication, 40*(1), 29–47.

Marcus, G. & Fischer, M. (1986). *Anthropology as cultural critique: An experimental moment in the human sciences*. Chicago: University of Chicago Press.

Opsal, T., Wolgemut, J., Cross, J., Kaanta, T., Dickmann, E., Colomer, S., & Erdil-Moody, Z. (2016). "There are no known benefits . . .": Considering the risk/benefit ratio of qualitative research. *Qualitative Health Research, 26*(8), 1137–1150.

Sinding, C. & Aronson, J. (2003). Exposing failures, unsettling accommodations: Tensions in research practice. *Qualitative Research, 3* (1), 95–117.

Wagner, I. & Wodak, R. (2006). Performing success: Identifying strategies of self-presentation in women's biographical narratives. *Discourse & Society, 17*(3), 385–441.

Weick, K. E. (1995). *Sensemaking in organizations*. Thousand Oaks, CA: Sage.

Lorusso, Mick. (2009). *Pulse.* Urban Cosmos Series, Meditative Interventions.
Found object photograph. 22 x 30cm.

Lorusso, Mick. (2010). *Emergent World*. Essence of Light/Life Series, Energy Patterns.
Drawing. Watercolor pencil on paper. 77 x 55 cm.

Cybernetics and Human Knowing. Vol. 28 (2021), nos. 3-4, pp. 33–47

Anthropology in the Shadow of Anthropocene Overheating
Pantheistic Atheism and the Biosemiotic Turn

Thomas Hylland Eriksen[1]

This article is about pattern resemblances and connections, and connects the dots between Bateson *père* and Bateson *fille*, then argues that a biosemiotic turn is long overdue in anthropology, and finally shows how the mode of thinking developed in Mary Catherine Bateson's work can shed light on global crises today.

The article is framed as a dialogue between anthropology and biosemiotics. In anthropology, there is currently a strong concern with overcoming the Cartesian dualisms and to include the non-human world in its theoretical framework; in biosemiotics, there is a need to show its relevance for the study of human relations. In this, the two are complementary, and whereas anthropology produces substantial knowledge about human diversity and communication, biosemiotics offers a methodology for studying communication at a multispecies level without relinquishing scientific standards or falling into the trap of anthropomorphism. Mary Catherine Bateson's relational and processual perspective on human lives contributes some of the tools needed to make sense of systems of a very different kind, notably global climate change and the effects of globalization on diversity.

Keywords: Biosemiotics, anthropology, diversity, flexibility, globalization

> For the great irony of our time is that, even as we are living longer, we are thinking shorter. —Mary Catherine Bateson, *Composing a Further Life.*

Like her parents, Mary Catherine Bateson was an anthropologist. Unlike her mother, however, she spent her life on the outskirts of the discipline and beyond it, making her name as an educator and theorist of learning, a scholar of language and metaphors, a biographer of her parents and a sensitive and wise writer about gender, generations and race. To many, she is best known as a co-author, with her father, of *Angels Fear*, an exploration of the sacred from a non-religious perspective. However, non-religious may not be the best designation of the father-and-daughter project carried through by Mary Catherine after Gregory's death in 1980. Rather, the book brims with religious notions, it buzzes with powerful metaphors spinning webs between all kinds of living things, and at the end of the day, perhaps it is best described as a work of pantheistic atheism, or atheistic animism (although this designation fits the father better than the daughter). As in Gregory Bateson's previous work *Mind and Nature* (1979), the concept of mind is seen as a supra-individual aspect of the biosphere, creating those connections which make communication possible and information, famously defined in *Steps to an Ecology of Mind* (1972) as differences that make a difference, available

1. University of Oslo. Email: t.h.eriksen@sai.uio.no

to all perceiving entities, albeit to varying extent and degree. In a certain sense, all that exists in the world are differences.

Mary Catherine Bateson was a relational, processual thinker who thrived on the small, but complex surfaces of interpersonal relations, but whose relevance extends into other domains, as she was well aware—among other things, she contributed actively to the climate debate. The pattern resemblances between otherwise vastly different systems, a premise in cybernetic thinking and often identified by her father, suggests that her thinking about generations, communication and the sacred is not only relevant for the global challenges and crises of our time, but may add a dimension to our understanding of life, by showing, in almost everything she wrote, the power and potentials of metaphorical thinking. As she asks in her penultimate book: "How is a mountain range like the sea? How is a park model trailer like a ship? How is the fashioning of delicate jewelry like the maintenance of great diesel engines?" (M. C. Bateson, 2010, p. 35). It is precisely by making these kinds of connections that it becomes possible to see not only differences that make a differences, but also similarities that create commonalities, often in surprising places. An additional quality of her work, which is so pervasive as to qualify as a red thread, lies in M. C. Bateson's interest in social and cultural change and its implications for learning, knowledge and being in the world. In this domain, the influence from both her parents is evident. In *New Lives for Old,* Margaret Mead (1956) explored cultural change in Melanesia, arguing that whereas some aspects of culture change rapidly, others change much more slowly if at all, often creating a mismatch or disintegration between different parts of the human life-world or *Umwelt*. We readily adopt technology and consumption patterns that make life easier, but changing kinship systems or work practices is a different matter. On his part, Gregory Bateson wrote extensively about learning, connecting it to the theory of logical types (he described *deutero-learning* as learning to learn) and asked questions about the ability of systems to learn, adjust and change. Since her early work—I have in mind especially her early personal report from the 1968 Burg-Wartenstein conference (M.C. Bateson, 1972), the influence from cybernetics has been evident throughout. The logical errors producing double binds, fundamental in Batesonian family therapy, as well as the runaway intensifications of schismogenic culs-de-sac, proved to be highly productive in M. C. Bateson's work on learning and age, which conveys not only cleverness and insight, but wisdom. In her article "Joint Performance Across Cultures: Improvisation in a Persian Garden" (M. C. Bateson, 1993), she shows how mutual understanding emerges not through translation and explanation, but by way of improvisation and emotional co-presence. She also shows how teachers must sometimes take on the part of pupils, since learning has to be a two-way process. Although by her own confession no Anna Freud, who dedicated her life to her father's legacy, there is a case to be made for Mary Catherine being one of the Platos her father, a sometimes gnomic Socrates, needed for his ideas to be unpacked and to flourish.

I never got to know Mary Catherine personally, and so I cannot tell to what extent she felt alienated from, or even indifferent to, trends and tendencies in mainstream sociocultural anthropology. Her most influential work has been outside anthropology narrowly defined, and her early work on Arab culture and Arabic language was nearly absent from her retrospective collection of essays, *Willing to Learn* (M. C. Bateson, 2004; cf., M. C. Bateson, 2010, p. 239). It is nevertheless exceedingly likely that recent intimations of a consistently ecological, relational anthropology would have pleased her. I shall now, tentatively, outline an option for an anthropology which tries to leave anthropocentrism behind without relinquishing scientific ideals, and which is determined to be relevant as an intellectual toolbox, aiming not only to understand the human condition, but also the world. As M. C. Bateson herself says, "double binds are by no means limited to the families of schizophrenics and indeed they may be characteristic of all multiply coupled and embedded systems such as we discover in the natural world" (M. C. Bateson, 2008, p. 23).

Anthropology and the World

Anthropology has always been informed and inspired by events and current concerns—one may only think of the pandemic or the Syrian refugee crisis for recent examples. The considerable interest in ethnicity and nationalism towards the end of the last century was a result of the very noticeable, and to some disappointing, shift from class politics to identity politics across the world; slightly earlier, feminism produced a heightened awareness of gender in the discipline, and historical processes such as the marginalization of indigenous groups and the aftermath of the Second World War, which saw the advent of decolonization, the civil rights movement and the Cold War, stimulated important work among anthropologists keen to understand not only what it is to be human, but also what the contemporary world is like, perhaps motivated by a desire to use knowledge to make the world safe for difference, to quote Ruth Benedict, "less unequal and saner." Currently, this underlying, and occasionally explicit, normative concern is expanding its scope beyond human bigotry and inequality to include the wider ecology, thereby responding to contemporary crises of climate and the environment.

In the present decade, the towering concerns informing the social sciences and humanities are to do with effects of the accelerated change of global neoliberalism on livelihoods, climate and the environment, and the concept of the Anthropocene suddenly seems to be everywhere. As is by now well known, the term was initially proposed by the atmospheric chemist Paul Crutzen (with Eugene Stoermer), who is also the co-author of a much cited article, co-written with his colleague Will Steffen and the historian John McNeill (Steffen, Crutzen & McNeill, 2007) on social implications of climate change.

The current popularity of the concept does not merely signal an increased engagement with climate and the environment, but also a view of human life as being planetary in its entanglements and seamlessly integrated with those of other species. In this shift lies a radical potential for rethinking what we do as anthropologists, whether we mainly try to understand the world or the human condition. Many scholars in the social sciences and humanities, whether or not they approve of the term Anthropocene (some prefer the *Capitalocene* or, in the case of Donna Haraway [2016], the *Chthulucene*), grapple with this shift, trying to reshape their disciplines to come to terms with what some speak of as the more-than-human world inhabited mainly by non-humans. These are superfluous terms of course—everybody who has heard about ecology knows that it is not exactly a groundbreaking discovery that most of the living world is non-human—but such earnest attempts to change dominant terminology are symptomatic of a deeper systemic change in the prevalent ecology of ideas and of practices. It may even retrospectively be observed that academics and intellectuals of the 2020s were in the midst of an intellectual revolution, only dimly aware of its long-term consequences—adding new dimensions to the *anthropos* of anthropology, expanding the humanities to include life in general, and recognizing that the relationships studied by the social sciences include social animals which are not human.

In this article, I propose a methodology for research on the ecological embeddedness of human lives which could be exactly what sociocultural anthropology needs in order to avoid the infinite regress of hardcore relativism, on the one hand, and the severe limitations of nature/culture dualism, on the other hand. Drawing on biosemiotics, and under inspiration from the Batesons, I shall eventually outline an approach where living systems are studied as systems of communication, a methodology which dissolves the nature/culture boundary without denigrating human agency, and which also has considerable comparative potential. Although biosemiotics has great potential as an epistemological and methodological framework for post-Cartesian research on human life-worlds, it is yet to become part of the standard toolbox of anthropological research. It is time for this to change. Much of what I have to say is familiar to most readers, but not necessarily the same bits. Hopefully, both biosemioticians and anthropologists can get something out of this, if not necessarily the same things.

Biosemiotics represents a way of talking about relationships among living things which avoids the pitfalls of Romantic essentialism, reductionist science and fundamentalist religion. Warning against the latter two, M. C. Bateson claims that the

> general public is, in a curious way, buying into a form of biological fundamentalism that is itself dangerous because of the metaphors of unilateral control that it proposes. Overemphasis on "master molecules" and "selfish genes" is as likely to lead to authoritarianism as is monotheism. (Bateson, 2008, p. 24)

Flexibility, seen as uncommitted potential for change, is being reduced in the contemporary world owing to the global commitment to economic growth, which entails standardization and leads to a loss of both biodiversity and cultural diversity; its corollary in the world of ideas is an overly confident reliance on a sparsely equipped toolbox.

Anthropology and Climate Change

As suggested, current attempts to transcend Cartesian dualism are motivated by the dramatic environmental changes wrought by human expansion, short temporal horizon and technological change. In recent years, the subject of paramount importance has been climate change. The body of knowledge that anthropologists have so far accumulated in this area ranges from critical studies of the discourses and practices of carbon offsets to comparative studies of retreating glaciers, in addition to a fast growing number of ethnographies describing how communities deal with the local effects of climate change, in projects that look, in Kirsten Hastrup's evocative terms, at the drying lands, the rising seas and the melting ice (Hastrup, 2009). A political economy approach informed by Marxist theory and anthropological reflexivity is provided, inter alia, in works by Hal Wilhite (2016) and Alf Hornborg (2019). Local responses to climate change are explored in Stensrud and Eriksen (2019; see Eriksen, 2016) and in Hoffman, Eriksen and Mendes (2021); the relationship between health, capitalism and climate has been analyzed by Hans Baer and Merrill Singer (2018), and the historical antecedents of current concerns with environmental change and climate are covered in Michael Dove's (2013) historical reader. Anthropologists have also contributed some very significant ethnographic monographs on climate issues, ranging from Jessica Barnes's (2014) research on water in the Nile delta to Linda Connor's (2016) work on mining in Australia, with Crate and Nuttall's *Anthropology and Climate Change*, from 2009, as an early benchmark collection. These are just a handful of examples illustrating a shift in the collective attention of anthropologists (see Fiske et al., 2014; Eriksen, 2021 for more comprehensive overviews).

Not all environmental anthropology has a focus on climate. Important research on topics such as deforestation, mining, waste and toxins may be only indirectly related to climate. However, it is fair to say that the broader field of environmental anthropology is being renewed and reformulated owing to the intensified attention to climate, as witnessed, for example, in the edited volume *The Angry Earth: Disasters in Anthropological Perspective* where, in the second, revised and updated edition of the book, nearly all contributors mention the atmospheric changes that have begun to affect the sites of their prior studies (Oliver-Smith & Hoffman, 2020). It also deserves mentioning that the most famous living anthropologist without an anthropology degree, Bruno Latour, shifted his attention years ago to the causes and politics of climate change (see e.g., Latour, 2017). It is everywhere, and it is now, and indeed, the fact of anthropogenic climate change may well be said to redefine not only the specialty of anthropological (or other) research, but raises the question of what it entails to be a human being within a new existential and conceptual framework. Climate change, the immediate cause of the coining of the neologism Anthropocene, may retrospectively be seen as a major game-changer in intellectual and political life in general, and also in anthropological research. It is no coincidence that the increased interest in multispecies fieldwork, and the rise to prominence of the Bateson-inspired term *assemblage* (which transcends the human–nonhuman and material–symbolic barriers; Deleuze & Guattari, 1980), have shaped the work of many anthropologists in the present century.

This leads up to the central question I am raising on this occasion, namely how research by anthropologists, who specialize in human relations, can extend beyond anthropocentrism and fully incorporate the biosphere or the ecosystems of which we are part.

It would be wrong to state that ecological thinking has never influenced anthropology in the past. In the mid 20th century, some American anthropologists, who traced their genealogy to the Victorian materialism of Lewis Henry Morgan rather than the humanities approach favored by Franz Boas and his students (among them Margaret Mead), began to study human societies in their ecological settings. Their social evolutionism sometimes led to research on ecological adaptation; an old book still sitting on my shelf from my undergraduate years is *Man in Adaptation* (Cohen, 1971), a reader analyzing the interaction between sociocultural forms and the environment in small-scale societies. Others explained social configurations through the interaction of technological and environmental factors, most starkly formulated in Leslie White's (1949) formula according to which a measure of social evolution was the amount of energy a society was capable of harnessing from its environment. This approach was popularized by Marvin Harris (1978), about whose riddles of culture it was nevertheless said, by the symbolist Marshall Sahlins, that formulating riddles about culture is an excellent idea—but it is somewhat disconcerting that the answer in Harris's case always seemed to be protein. These approaches went out of fashion long ago, while other forms of ecological thought continue to inspire research.

A Relational Alternative to Materialism

Foremost of the less deterministic approaches, which never challenged the basic distinction between culture and nature, was Gregory Bateson and his monism. His ecological reading of the human condition was equally informed by his undergraduate training as a biologist, his interest in signs and communication, and his early fieldwork experience in New Guinea. Bateson's somewhat scant and scattered writings, most famously compiled in the uneven, but pathbreaking *Steps to an Ecology of Mind* (1972), suggest that his best ideas are far from dated, that they may indeed be seen as more acutely relevant today than in the mid 20th century when they were initially formulated. In Mary Catherine's work on metaphors, learning and aging, and in her editing and completion of *Angels Fear*, they are fleshed out and made relevant in new ways.

Although he was trained as an anthropologist, Gregory Bateson's initial intention was to become a biologist like his father William Bateson. In fact, he was converted to anthropology on a train to Cambridge in the company of Alfred Haddon, another natural scientist who had made the shift to anthropology (Lipset, 1982). In his *Anthropology and Anthropologists*, Adam Kuper (1996) remarks that Bateson did not quite fit into 1930s British anthropology, at the time dominated by the towering figures of Radcliffe-Brown and Malinowski, separated by their opposing views on structure and the individual, but united in their concern with the mechanisms of social integration and indifference to ecology.

Bateson was different, given his consistent interest in relationship, process and the logic of living systems (Wardle, 1999). He would soon move on to cybernetics, of which he was among the founders, psychiatry and general systems theory. Among his most powerful concepts are those of schismogenesis (self-reinforcing, usually destructive relationships), flexibility (uncommitted potential for change) and double-bind (irresolvable dilemmas resulting from errors in communication). Developing Bertrand Russell's theory of logical types, which states that a class cannot be a member of itself since it exists at another logical level, he wrote about meta-communication among humans and animals. A dog playing with another dog, or with a human for that matter, may display the same kind of aggressive behavior as a dog intent on attacking and inflicting injury, but by wagging its tail and only pretending to bite, it sends off the meta-message that it is just pretending. In later work on dolphin communication, Bateson similarly looked for logical types and different registers of communicating.

As a contributor to biosemiotics *avant la lettre*, Bateson is essential. In "Cybernetics of the Self" (in *Steps*), he shows that if an alcoholic takes recourse to willpower as a means to stop drinking, he will fail, since his problem is relational and systemic. The alcoholic thus has to give up his erroneous epistemology according to which he is the captain of his soul, accept that he is a part of something larger than himself, dependent and entangled.

Explaining the epistemological difference between individualism and a cybernetic view of an action, Bateson goes on to write:

> Consider a man felling a tree with an axe. Each stroke of the axe is modified or corrected, according to the shape of the cut face of the tree left by the previous stroke. This self-corrective (i.e., mental) process is brought about by a total system, tree-eyes-brain-muscles-axe-stroke-tree; and it is this total system that has the characteristics of immanent mind.
>
> More correctly, we should spell the matter out as: (differences in tree) – (differences in retina) – (differences in brain) – (differences in muscles) – (differences in movement of axe) – (differences in tree), etc. What is transmitted around the circuit is transforms of differences. And, as noted above, a difference which makes a difference in an idea or unit of information.
>
> But this is not how the average Occidental sees the event-sequence of tree-felling. He says, "I cut down the tree" and he even believes that there is a delimited agent, the "self," which performed a delimited "purposive" action upon a delimited object. (Bateson, 1972, pp. 444–445)

Through his refusal to distinguish between the material and the immaterial (the tree and the axe belong to the same system of signification as the man's intentions), Bateson successfully transcended Cartesian dualism. This radical move, familiar to Batesonians and biosemioticians, could have important implications for methodology in an anthropology determined to move beyond the mere *anthropos* as its empirical focus.

The Batesonian concept of mind, developed in *Mind and Nature* (1979) and elaborated in *Angels Fear* (Bateson & Bateson, 1988) is an important condition for this endeavour to be possible: It does not end at the skin, but is a property of the living systems of which you and I form part. The fiction of the bounded individual with their exclusive, limited mind can be challenged from many directions, not all of them ecological. Medical scholars may point to our reliance on the human microbiota, the millions of bacteria existing in a symbiotic relationship with the human organism, while cognitive scientists have shown that most of what people think they know is in fact collective knowledge, most of which exists outside our individual minds (Sloman & Fernbach, 2017). No individual has adequate knowledge of the 30,000 parts that make up a Toyota car, and yet these cars are assembled without a fault every day on the assembly lines in the vast factories of Toyota City. Peirce argued against Cartesian dualism and the atomistic *cogito* in the latter decades of the 19th century, showing that systems of signification were by definition shared and relational, with profound implications for ideas of personhood and mind.

The concept of the *Umwelt* is an additional tool enabling a liberation from dualism, and unlike Bateson's example of the man chopping down a tree, von Uexküll's older concept brings back the intentional subject and forms a bridge to phenomenology (Schroer, 2021; Tønnessen, 2021). The Umwelt, referring to an organism's perceptions of the environment—what we might today, using phenomenological language, call structures of relevance—is enabled and constrained by the organism's immediate interests, usually food and sex, and the limitations of its sensory apparatus. This perspective differs from standard Darwinian biology through its insistence on communication and signification as fundamental in nature; from ethology through an almost ethnographic approach to animals, trying to describe their activities in their own terms. Unfortunately, von Uexküll rejected mainstream evolutionary theory, and at least partly for this reason, he is all but absent from standard accounts of the history of biology (Schroer, 2021).

The Problem of Anthropocentrism

There are some fundamental problems with interpretive descriptions of animal behavior, the most obvious one being anthropocentrism. The social anthropologist Edwin Ardener (1989), examining ethological descriptions of animal behavior and demonstrating that designations of identical behaviors by different scholars varied wildly, concluded that the tendency to attribute human character traits to animals was a kind of scientific totemism. Claude Lévi-Strauss (1985), inverting his theory of totemism, once reflected on the attribution of presumed animal-like characteristics to humans, even showing pictures of men allegedly resembling the animals their personalities were like (such as bears, pigs and dogs). Claiming that a person is cunning as a fox, or sly as a snake, is a kind of second-order anthropomorphism. Allowing people and animals to be radically different is not easy. About the celebrated chimpanzee Nim Chimpsky, who was able to use more than a hundred signs, it may well be said that what his achievement really tells us is that apes are not particularly good at being human. Measuring chimpanzee intelligence with standards taken from human intelligence does not teach us much about chimpanzees, but perhaps a little about human foibles.

A much cited philosophical paper about humans and animals is Thomas Nagel's "What Is it Like to Be a bat?" (Nagel, 1974). The article is really about consciousness, but bats figure prominently because of their outlandish habits (to humans)—they use echolocation, hang upside down, and so forth. Nagel's view is that it is impossible for humans to know how being a bat is experienced. Knowledge is always positioned, and as natural beings, not supernatural observers describing the world "from nowhere," as Nagel has phrased it. We are embedded and constrained by our evolutionary history and sensory apparatus, like all other organisms. Knowing another organism's inner life is akin to knowing the *Ding an sich* independently of the limitations of our perceptual apparatus, which is impossible. Knowing what goes on inside another organism is impossible. What we can do, following the cues from the pioneers of biosemiotics, is to describe that which goes on between organisms. Von Uexküll's distinction between *Umwelt* and *Innenwelt* is helpful here, but it is ultimately insufficient, as it stops at the subjectively perceived world-out-there (Umwelt), not granting the semiosis autopoietically produced in the environment a distinct identity. This is precisely the point where contemporary biosemiotics moves beyond von Uexküll's Umwelt.

So how can we develop a non-anthropocentric methodology and conceptual world for observing and describing animal behavior as well as animal–human relationships?

Semiotic Freedom

In a bid to propose a methodology focusing on relationship and process rather than individual perceptions, I take my cue from one of the central theorists in contemporary biosemiotics, the late Danish thinker Jesper Hoffmeyer (1941–2019), whose concept of semiotic freedom is a fruitful starting-point. Hoffmeyer had a suitably interdisciplinary background in chemistry, biology, philosophy and semiotics. In biosemiotics, relationships in nature are interpreted as acts of communication. When a fox becomes aware of a hare in the vicinity, its reaction forms part of a semiotic chain together with the hare's response and flight, the hunt and its outcome. Hoffmeyer once said that if he were to summarize the entire history of evolution in one sentence, he would say that evolution has, over the millions of years, led to an overall growth in semiotic freedom (Hoffmeyer, 1998). The concept is proposed as a way of distinguishing the scope and range of communicative possibilities available to different species, aided by various degrees of semiotic scaffolding, that is, cues in the environment that increase the semiotic freedom of organisms, but also made possible thanks to the flexibility of the genes. Similar to stimuli humans get at academic conferences.

A dog has greater semiotic freedom than a snake, since it can anticipate the movement of a mouse that it is chasing even when the mouse has found a hiding place rendering it temporarily invisible. As humans, we may understand the dog's hunting strategy, but only up to a point. Seen with Nagel's criteria of consciousness, it is more alien to us than even the most impenetrable cosmology found in nonliterate societies. It brings to mind Wittgenstein's famous observation that "if a lion could speak, we would not understand him" (Wittgenstein, 1953/1983, p. 221). Conversely, it is unlikely that a lion would understand what a human meant—even if it could make sense of the words—when the human says that "my father was poor, but at least he was honest."

The fraught relationship between biosemiotics and Darwinism goes back to von Uexküll's anti-Darwinism, and given the recent predominance of selfish-gene Darwinism, it is understandable. However, Darwin himself was more of an ecological thinker than most of his later supporters. In the final sections of *Origin of Species*, Darwin speculates that when humans rank natural species, they come out on top because they place (cognitive) intelligence above other qualities, but if a bee were to do the same thing, it were likely to mention some instinct. In the very last paragraph of *Origin*, the Victorian biologist seems to intimate that the most important implication of his theory is the fact of interconnectedness of all that lives:

> There is grandeur in this view of life, with its several powers, having been originally breathed into a few forms or into one; and that, whilst this planet has gone cycling on according to the fixed law of gravity, from so simple a beginning endless forms most beautiful and most wonderful have been, and are being, evolved. (Darwin, 1859, p. 490)

In earlier chapters, Darwin devotes many pages and a great deal of enthusiasm to the wonders of the geometrically perfect beehive and the conundrum of the sterile working-ant. Although he framed his findings in the language of mid nineteenth-century English liberalism, emphasizing competition rather than complementarity, the first edition of *Origin* is not a story about progress, but one about shared origins and the way in which all life is ultimately one. A reading of *Origin* along these lines (bracketing for now his later work, written under the influence of contemporaries such as his cousin Francis Galton and Herbert Spencer) would reveal an ecological evolutionist rather than an individualist one marinated in Whig ideology.

Darwin, thus, was on the brink of pitching his doctrine of natural selection as an ecological vision rather than one founded in competition and "nature red in tooth and claw." Peirce's and von Uexküll's contributions can, in this light, be seen as elaborations rather than competing views. As Darwin's comparison between humans and bees may indicate, grading species on a scale of semiotic freedom is not interesting; the focus should rather be on the fact that reading relationships and processes in nature as signs and communicative networks can provide a recipe for serious multispecies or ecosystemic work. It is a matter of some interest that Peirce, the founder of semiotics, was a philosopher and logician, whose theory of signs influenced biologists like Hoffmeyer, whose insights into the differences that make a difference in nonhuman nature can now make a return to the study of human communication, but with an important twist, namely a seamless integration with other living organisms through the lens of semiosis, where difference is a matter of degree made visible through relationships rather than one of kind, made visible through the designation of boundaries. Semiotic scaffolding is provided by the *Umwelt*, which is less stable and more volatile than imagined by von Uexküll (Tønnessen, 2009). As Donald Favareau explains on the International Society of Biosemiotics (ISBS) website:

> Sign relations make possible not only such higher-order human abilities as spoken language and written texts, but also such communicative animal behavior as the calls and songs of birds and cetaceans; the pheromone trails of insect colony organization and interaction; the mating, territorial, and hierarchical display behavior in mammals; as well as the deceptive scents, textures, movements and coloration of a wide variety of symbiotically interacting insects, animals and plants. (quoted in Favareau, 2015, p. 229)

A particularly intriguing text for the effort to contribute to a biosemiotic turn in anthropology is Eduardo Kohn's *How Forests Think* (2013), which simultaneously draws on biosemiotics, the native's point of view and conventional ecological anthropology. Interestingly, Kohn's supervisor was Terrence Deacon, a biological anthropologist inspired by Peirce and Bateson, and whose *Incomplete Nature* (2012), a landmark contribution to biosemiotics, restricts its case material mainly to the non-human world. Kohn's book, by contrast, is an ethnography of the Runa in Ecuador and their engagement with their environment, which has interesting parallels in the biosemiotic approach. According to the Runa logic, since rain starts when ants appear, people are able to impede rain by smoking tobacco, as the smoke prevents ants from coming out. This is obviously nonsense, but in their world, this kind of causality makes sense. The similarity with biosemiotics consists in the view that the world is being reproduced through communication or semiosis. While animists or panpsychists conceive of the world as being imbued with an immanent intentionality or mind, biosemioticians regard natural processes as systems of signification. Regardless of deeper ontological differences, there is a convergence here, the main difference being that biosemiotics is committed to scientific procedures of validation and falsification. There is no reason to assume, or demand, that Runa or other indigenous groups should embrace biosemiotics at the expense of their traditional ways of engaging with their worlds. After all, they are not academics, and science enters into their lives only peripherally. However, as the philosopher Arne Johan Vetlesen (2019) argues in an exploration of panpsychism in light of the environmental crisis, there are ways of relating to the human Umwelt which would prevent the massive violence and destruction resulting from modernity, but which have partly been forgotten and might be recovered. It is striking how the vast majority of stateless societies regard themselves as an integral and integrated part of their ecological surroundings, while moderns draw a boundary. In other words, the non-Cartesian ontologies predominate in cultural history, and there are good reasons to recover their ethos for the sake of the planet, but also for a scientific endeavour that asks a new kind of question.

Allow me now to return to the concept of semiotic freedom and its implications for an anthropology of the present. The concept, and the entire biosemiotic approach, has some obvious advantages compared to its alternatives.

First, the problem of subjectivity or intentionality, which is a classic problem in cybernetics, where conscious purpose tends to be bracketed, vanishes. Semiotic freedom indicates the potential to act otherwise, whether the organism in question is a beetle or a cat, but it is graded. One unanswered question, inherited from Peirce, is how something becomes someone. By this he has intentionality, directedness of consciousness or will in mind. This is not the place to begin to answer that question, which is unlikely to go away any time soon.

Non-vertebrates, Hoffmeyer assumes, act mainly on the basis of instinct, and therefore the term *gift* is unfortunate when used about apparent reciprocity between insects during courtship and mating. Vertebrates, and especially mammals, are forced to take decisions at the spur of a moment; whether to pounce or wait, fight or flee, go right or left; their capabilities are shared throughout the species, but their inherited affordances do not give detailed instructions as to how to act in a given situation. In this sense, antelopes and cats are like you and me—they have to decide. Yet the similarities should not be overestimated, and an interspecies comparison will quickly reveal an uneven distribution of semiotic freedom. The ability of metacommunication is more easily observed in apes and dogs than in sheep and crocodiles; and recall that even the ape genius Nim Chimpsky developed a repertoire of 125 words in American Sign Language, less than a normal human one-year old. In other words, differences that make a difference continue to exist in a continuous world, but mind, in Bateson's sense in *Mind and Nature*, is distributive. Organisms do not stop at the skin, but at the frontier of their Umwelt.

Hoffmeyer has written: "Semiotic freedom may in fact be singled out as the only parameter that beyond any doubt has exhibited an increasing tendency throughout the evolutionary process" (Hoffmeyer, 2008, p. 39). There is more semiosis, more communication, more complexity in the natural world than ever, notwithstanding its temporary reduction owing to mass extinctions in the past.

However, it may be argued that semiotic freedom is now being reduced owing to pollution, species extinction, habitat loss, captivity, plantation-like standardization and simplification of ecosystems. Thus, biosemiotics can be a tool for critiquing the homogenizing effects of global capitalism.

The concept of semiotic freedom, moreover, confirms the relevance of the term *Anthropocene*, criticized by some as being overly anthropocentric. However, not only are humans responsible for bringing about the ecological crisis, but humans also have far greater semiotic freedom, in some key domains, than any other species.

In this actually existing world, we human beings have special responsibilities, and this is not just because we possess, and make use of, weapons of mass destruction, from the steam engine to the mobile telephone, but because of our superior semiotic freedom. While the semiotic freedom of a snake is limited by its sensory apparatus—it stops chasing a mouse when it can no longer sense it—a dog will be aware of the mouse's existence even when it is hiding in a heap of rotten leaves. A human, furthermore, will not only know of the mouse, but will also be capable of predicting the future growth of rodent populations and improve their mousetraps. The difference is one of degree, but it makes a great deal of difference. If evolution, as Hoffmeyer would have it, consists in the growth in semiotic freedom (a stimulating way of talking about complexity), then we, *homo sapiens*, for the time being placed at the pinnacle of this process, are the only species capable of reducing the amount of semiotic freedom in the natural world, and we are well on our way to doing so.

Finally, if the Batesons, as they come across in *Angels Fear*, could be considered a kind of pantheistic atheists, and I think they can (bracketing for the moment Mary Catherine's religious turn), then biosemiotics could be seen as a form of scientific animism—recall the Runa worldview and its connections with biosemiotics. Although it is true that the standard scientific tradition is dominated by a misleading and unhelpful Cartesian dualism, there have always been dissenting voices and alternative methodologies, although few have passed the test of time. Goethe was an early advocate for a qualitative natural science, and interpretive methods in biology have been brought to fruition in exciting ways by the biosemiotics movement. It offers a solution to the problems encountered in attempts to carry out multispecies fieldwork, and now that the Anthropocene moment redefines priorities and paradigms, biosemiotics can open doors that would otherwise have remained blind or locked. Such an achievement would be in the spirit of Mary Catherine Bateson, who concluded a revisit to *Angels Fear* by stating that "thinking in terms of systems offers a … kind of holism where we can see the similarities between ourselves and systems of many kinds, not only organisms but ecosystems and human communities, and we can see them living, responding, and changing" (M. C. Bateson, 2008, p. 24). Although her work focused mainly on relationships between people, the vision behind assumed that otherwise very different systems display many of the same characteristics. Phrased in the right way, and avoiding the pitfalls of conflating logical types, it may be shown that the shifting meaning of age—which is a relational concept par excellence—can tell us something necessary about relationships in the natural world and the challenges from Anthropocene effects.

Acknowledgement

The author would like to thank Fred Steier for his comments on the first draft.

References

Ardener, E. (1989). *The voice of prophecy and other essays* (M. Chapman, Ed.). Oxford, UK: Blackwell.
Baer, H. A., & Singer, M. (2018). *The anthropology of climate change: An integrated critical perspective* (2nd ed.). London: Earthscan at Routledge.
Barnes, J. (2014) *Cultivating the Nile: The everyday politics of water in Egypt.* Durham, NC: Duke University Press.
Bateson, G. (1972). *Steps to an ecology of mind.* New York: Bantam.

Bateson, G. (1979). *Mind and nature: A necessary unity.* Glasgow: Fontana.
Bateson, G., & Bateson, M. C. (1988). *Angels fear. Towards an epistemology of the sacred.* Chicago: University of Chicago Press.
Bateson, M. C. (1993). Joint performance across cultures: Improvisation in a Persian garden. *Text and Performance Quarterly, 13,* 113–121.
Bateson, M. C. (2004). *Willing to learn. Passages of personal discovery.* Lebanon, NH: Steerforth.
Bateson, M. C. (2008). Angels fear revisited: Gregory Bateson's cybernetic theory of mind applied to religion-science debates. In J. Hoffmeyer (Ed.), *A legacy for living systems: Gregory Bateson as precursor to biosemiotics* (pp. 15–26). Springer.
Bateson, M. C. (2010). *Composing a further life: The age of active wisdom.* New York: Knopf.
Cohen, Y. (Ed.). (1971). *Man in adaptation: The institutional framework.* London: Taylor & Francis.
Connor, L. (2016). *Climate change and anthropos: Planet, people and places.* London: Routledge.
Crate, S. A. & Nuttall, M. (Eds.). (2009). *Anthropology and climate change: From encounters to actions.* Walnut Creek, CA: Left Coast Press.
Darwin, C. (1959). *On the origin of species by means of natural selection, or the preservation of favoured races in the struggle for life.* London: John Murray.
Deacon, T. 2012. *Incomplete nature: How mind emerged from matter.* New York: Norton.
Deleuze, G., & Guattari, F. (1980). *Mille plateaux.* Paris: Minuit.
Dove, M. R. (Ed.). (2013). *The anthropology of climate change: An historical reader.* Chichester: John Wiley & Sons.
Eriksen, T. H. (2016). *Overheating: An anthropology of accelerated change.* London: Pluto.
Eriksen, T. H. (2021). Climate change. *Cambridge Encyclopedia of Anthropology.* https://www.anthroencyclopedia.com/entry/climate-change
Favareau, D. (2015). Why this now? The conceptual and historical rationale behind the development of biosemiotics. *Green Letters: Studies in Ecocriticism, 19*(3), 227–242.
Fiske, S. J., Crate, S. A., Crumley, C. L., Galvin, K., Lazrus, H., Lucero, L. Oliver-Smith, A., Orlove, B., Strauss, S., Wilk, R. (2014). *Changing the atmosphere. anthropology and climate change.* Final report of the AAA Global Climate Change Task Force. December 2014. Arlington, VA: American Anthropological Association.
Haraway, D. (2016). *Staying with the trouble: Making kin in the Chthulucene.* Durham, NC: Duke University Press.
Harris, M. (1978). *Cannibals and kings: The origins of culture.* Glasgow: Fontana.
Hastrup, K. (2009). *The question of resilience: Social responses to climate change.* Copenhagen: The Royal Danish Academy of Sciences and Letters.
Hoffman, S. M., Eriksen, T. H., & Mendes, P. (Eds.). (2021). *Cooling down: Local responses to global climate change.* Oxford, UK: Berghahn.
Hoffmeyer, J. (1998). Semiosis and biohistory: A reply. *Semiotica, 120*(3/4), 455–482.
Hoffmeyer, J. (2008). From thing to relation. On Bateson's bioanthropology. In J. Hoffmeyer (Ed.), *A legacy for living systems: Gregory Bateson as precursor to biosemiotics* (pp. 27–44). Springer.
Hornborg, A. (2019). *Nature, society, and justice in the Anthropocene: Unraveling the money–technology–energy complex.* Cambridge, UK: Cambridge University Press.
Kohn, E. (2013). *How forests think: Toward an anthropology beyond the human.* Berkeley, CA: University of California Press.
Kuper, A. (1996). *Anthropology and anthropologists: The modern British School,* (2nd ed.). London: Routledge
Latour, B. (2017). *Down to Earth: Politics in the new climatic regime.* Cambridge, UK: Polity.
Lévi-Strauss, C. (1985). *La potière jalouse.* Paris: Plon.
Lipset, D. (1982). *Gregory Bateson. The legacy of a scientist.* Boston: Beacon Press.
McNeill, J. R. & Engelke, P. (2016). *The great acceleration.* Cambridge, MA: Harvard University Press.
Mead, M. (1956). *New lives for old – Cultural transformations, Manus 1928–1953.* New York: Harper Collins.
Nagel, T. (1974). What is it like to be a bat? *The Philosophical Review, 83*(4), 435–450.
Oliver-Smith, A. & Hoffman, S. M. (Eds.). (2020). *The angry Earth: Disaster in anthropological perspective* (2nd ed.). London: Routledge.
Schroer, S. A. (2021). Jakob von Uexküll: The concept of Umwelt and its potentials for an anthropology beyond the human. *Ethnos, 86*(1), 132–152
Sloman, S., & Fernbach, P. (2017). *The knowledge illusion: The myth of individual thought and the power of collective wisdom.* New York: Riverhead.
Steffen, W., Crutzen, P. J., & McNeill, J. R. (2007). The Anthropocene: Are human beings now overwhelming the forces of nature? *AMBIO, 36*(8), 614–621.
Stensrud, A. B., & Eriksen, T. H. (Eds.). (2019). *Climate, capitalism and communities.* London: Pluto.
Tønnessen, M. (2009). Umwelt transitions: Uexküll and environmental change, *Biosemiotics, 2*(1), 47–64.
Tønnessen, M. (2021). *The relevance of Umwelt theory for the theory and practice of phenomenology.* Working paper, University of Stavanger.
Vetlesen, A. J. (2019). *Cosmologies of the Anthropocene: Panpsychism, Animism, and the limits of posthumanism.* London; Routledge.
Wardle, H. (1999). Gregory Bateson's lost world: The anthropology of Haddon and Rivers continued and deflected. *Journal of the History of the Behavioral Sciences, 35*(4), 379–389.
White, L. (1949). *The science of culture: A study of man and civilization.* New York: Grove Press.

Wilhite, H. (2016). *The political economy of low carbon transformation: Breaking the habits of capitalism*. London: Routledge.

Wittgenstein, L. 1983. *Philosophical investigations*. Oxford, UK: Blackwell. (Originally published in 1953)

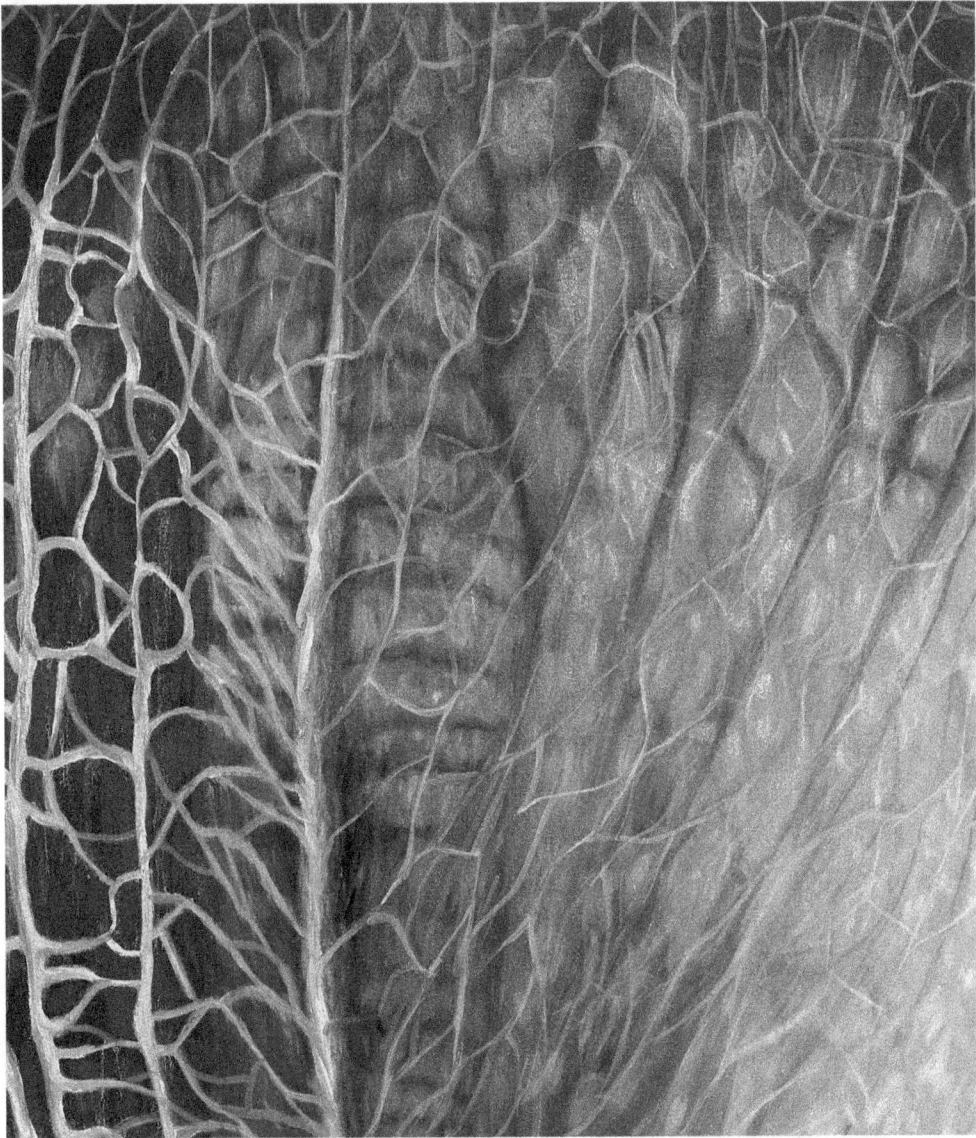

Lorusso, Mick. (2005). *Michael*. Cosecha de Luz Series, Energy Patterns.
Painting. Oil on canvas. 76 x 85 cm.

Lorusso, Mick. (2010). *Consolidate*. Essence of Light/Life Series, Energy Patterns.
Drawing. Watercolor pencil on paper. 77 x 55 cm.

Cybernetics and Human Knowing. Vol. 28 (2021), nos. 3-4, pp. 49–67

Interdependence Is the Key Issue:
Mary Catherine Bateson and
the Myth of Individualism

Alfonso Montuori[1]

This paper explores Mary Catherine Bateson's critique of individualism and argument for the recognition and development of interdependence. Individualism is presented as a cornerstone of American culture, forming a constellation along with analytic thinking, methodological individualism and free market capitalism. This constellation stands in opposition to and is held in place by the constellation of collectivism, holistic thinking, methodological holism and socialism/ communism. These constellations and their dimensions are viewed as static and reified and polarized to the extent that they are no longer useful. Creativity is introduced as a factor for change and the elements of process and transformation. The discussion is contextualized drawing on examples from both academia and popular culture.
Keywords: collectivism, creativity, genius, holism, individualism, interdependence, jazz

> The pressure toward postmodernism is building from our lack of ability to overcome certain dualisms that are built into modern ways of knowing. —Ogilvy (1989, p. 9).

> Total independence is an imaginary construct, a limit case of interdependence which is universal. —McGilchrist (2021, p. 35).

> We are all caught in an inescapable network of mutuality, tied into a single garment of destiny. Whatever affects one directly, affects all indirectly...before you finish eating breakfast in the morning, you've depended on more than half the world. This is the way our universe is structured, this is its interrelated quality... We aren't going to have peace on Earth until we recognize this basic fact of the interrelated structure of all reality. —Martin Luther King (Washington, 2003, p. 254)

The 20th century was a dramatic time for science and for humanity. Humanity's understanding of the universe changed in dramatic and mysterious ways (Peat, 2002). The physicist Paul Davies described the change in broad strokes:

> For three centuries, science has been dominated by the Newtonian and thermodynamic paradigms, which present the universe as either a sterile machine, or in a state of degeneration and decay. Now there is the paradigm of the creative universe, which recognizes the progressive, innovative character of physical processes. The new paradigm emphasizes the collective, cooperative, and organizational aspects of nature; its perspective is synthetic and holistic rather than analytic and reductionist. (Davies, 1989, p. 2).

Science may be moving forward in leaps and bounds, but that same 20th century saw, among other things, the horrors of two World Wars, and this paradigm of the creative

---

1.   California Institute of Integral Studies. Email: amontuori@ciis.edu

universe is struggling to make itself heard in the social sciences and in our everyday lives, where if anything there is more of a sense of postmodern degeneration and decay. The paradigm of the creative universe is radically different from Newton's clockwork universe. Davies outlines two fundamental differences that will be relevant in this article. They are the systemic and profoundly interconnected nature of a Universe that is fundamentally creative. What we might call the way of knowing that followed from the Newtonian view focused on parts rather than wholes and on order rather than the creative dialogic between order, disorder, and organization (Morin, 2008a). The Newtonian/Cartesian worldview saw human beings as machines, made up of parts, and as individuals who are separate from fellow humans and nature (Capra & Luisi, 2014). How might human beings view themselves in this new paradigm of creative interdependence? First they would have to untangle the complexities of the dualisms of modernity. This will be the heart of our inquiry, prompted by the writings of Mary Catherine Bateson.

One of Mary Catherine Bateson's last articles addressed the myths of independence and competition (Bateson, 2016). The themes of independence, individualism, competition and humanity's relationship with the environment were central to the work of both Gregory Bateson and Mary Catherine Bateson. These themes are of philosophical as well as practical interest. During a global time of transition, when one age is dying and a new one has not yet emerged, during a pandemic, when issues of individual freedom, independence, interdependence and arguably human identity itself have come to the forefront, reflecting on these cornerstones of human existence is more important than ever. They pertain directly to the emergence of the paradigm of a creative universe and arguably to human survival.

Mary Catherine (Bateson, 2015) asked: "Have we arrived at the point of looking at ourselves as parts of larger systems on which we depend?" (p. 36). She went on to say that "interdependence is really the key issue" (p. 36). If modernity and the Newtonian/Cartesian paradigm involved the emergence of the individual and individualism in the history of the West (Shanahan, 1992; Taylor, 1992), then with the emergence of the Anthropocene—a geological age named after humanity's influence on climate and the environment—there is a need to reconsider these "larger systems on which we depend" (Bateson, p. 36).

Here is Gregory Bateson on epistemology and the importance of how human beings see themselves, establish boundaries, and identify vis-à-vis their environment:

> When you narrow down your epistemology and act on the premise "What interests me is me, or my organization, or my species," you chop off consideration of other loops of the loop structure. You decide that you want to get rid of the by-products of human life and that Lake Erie will be a good place to put them. You forget that the eco-mental system called Lake Erie is part of your wider eco-mental system—and that if Lake Erie is driven insane, its insanity is incorporated in the larger system of your thought and experience. (G. Bateson, 1972, p. 484)

To have such an epistemology is to view oneself as what Bateson's contemporary, the philosopher Alan Watts, called a skin-encapsulated ego (Watts, 1989), with boundaries firmly drawn around one's skin, beyond which everything and everyone else is other. In his study of individualism in the United States, Traber (Traber, 2007) reminds us that "one of the nation's ruling myths continues to be that the self-contained individual is unconstrained by society, culture, and history" (p. 1). In what has become the ideology of individualism Traber describes, this is the view of the person as a closed system, gradually driving itself insane.

Mary Catherine was clear about one of the key obstacles to a more systemic view of the world: "I regard the United States as an extremely individualistic society, in which individualism is often a form of conformity and often, too, a justification of competition in preference to cooperation" (Bateson, 2016, p. 675).

Stewart and Bennett made some broad generalizations about North American cultural patterns that provide some relevant background on individualism (Stewart & Bennett, 1991). On the basis of cross-cultural comparisons they argued (a) that North Americans view the self as the "cultural quantum in society" (p. 134); (b) that "in the American self, there is a remarkable absence of community, tradition, and shared meaning which impinge upon perception and give shape to behavior" (p. 130); (c) that North Americans reject "sociological and philosophical principles" (p. 135) and replace them with psychological theories; (d) that the nature of North Americans' self-concept prevents them from understanding the enormous cross-cultural variations in self-concepts; and (e) that despite the emphasis on freedom of choice and autonomy, North Americans are subject to subtle but pervasive pressures to conform, to be free like everybody else.

Mary Catherine Bateson went on to say that "I think we all know that independence is an illusion. There is no such thing as independence, but it is a very powerful illusion that drives people's behavior, very often including anti-social behavior" (2016, p. 675).

This is a big claim, and it is a claim that reflects the systemic worldview articulated by Davies. But independence has become a term with heavy ideological, political, and cultural baggage, largely because of its association with individualism and freedom and. I believe "individual" for Mary Catherine means something closer to her father's statement that "the unit of survival is organism plus environment" (G. Bateson, 1972, p. 483). It is an illusion to think there is an individual independent of their environment, and yet this assumption is built into a certain kind of individualism. The historian Daniel Boorstin (Boorstin, 1965) wrote that "of all American myths, none is stronger than that of the loner moving west across the land" (p. 87). This myth of the self-reliant lone individual fighting against all obstacles, environmental and human, is at the root of much of American individualism (Bellah, Madsen, Sullivan, Swidler, & Tipton, 1985). It persists even in the networked age, when most American loners can be unimaginably more connected by virtue of having a computer than the most connected American in the age "when the West was won." Yet Boorstin reminded his readers that the reality of the frontier involved groups of travelers, and that cooperation and community were in fact essential for survival.

To make the Batesons' claim about independence is to go up against an interrelated network of ideas that has become a political ideology. As a result, the pushback against challenges to this myth is considerable because the myth is perceived to be at the heart of a certain understanding of American national identity. For some, to question it is to question America and what it means to be an American. In the next section I will briefly discuss my own experience with a critique of the hyper-individualist interpretation of creativity which sees it as the exclusive result of a lone genius to outline some of the key dimensions of this individualist ideology. I will show how the lone genius myth obscured other forms of creativity and suggest that it is in fact these very same relational, networked forms of creativity that might help go beyond the pernicious polarity of individualism versus collectivism.

The Lone Genius

Long ago and far away in London, England, I was a professional musician. I grew up listening to bands, I played in bands, and to this day I still play and perform in bands. When I started exploring creativity research in the mid-80s I was naturally curious to see what creativity research had to say about bands. To my surprise there was hardly anything about creative groups, musical or otherwise. There was nothing about creative relationships, or about environments that supported or inhibited creativity. With a few exceptions, creativity research was focused on the creative individual. The research was loosely organized around what were known in creativity research as the three P's: Person, Product, and Process (Runco, 2007). The who of creativity was therefore clearly established: it was a person, not a band or a relationship. The research was on writers, composers, poets, creators who could be viewed as laboring in isolation, but hardly ever performing artists. Why was there no research on groups, I wondered? I became fascinated by this question. Tellingly, there was research on groupthink (Janis, 1983), on the dangers of conformity in groups, but not on relational, social, group creativity. The fact that there was research on groupthink reflected the view that, if anything, others stood in the way of the creative individuals and groups led to conformity. Other people were mostly obstacles, not sources of inspiration and collaboration. How could it be that nobody had thought of researching more relational creativity until the 80s, and that there are still scholarly articles wondering whether group or collective creativity is an oxymoron (Chaharbaghi & Cripps, 2007; Staw, 2009)?

My colleague Ron Purser and I wrote an article in which we argued for a more relational and contextual view of creativity (Montuori & Purser, 1995). We argued that the lone genius was not so alone, but part of a larger system which had contributed language, concepts, tools, traditions (musical, scientific, visual), cultures, as well as friendships, colleagues, benefactors, craftspeople, critics, and competitors. We also argued for the importance of studying groups and environments that support or inhibit creativity, and that the exclusive focus on the individual genius prevented researchers and popular culture in general from seeing other manifestations of creativity. We were essentially proposing a systemic, contextual, relational view of creativity. In such a view, individuals are considered open systems, in constant exchange with their environment, rather than closed systems, as had been the tradition in research on the psychology of creativity, with notable exceptions (Barron, 1995). As Mary Catherine Bateson put it, we saw individuals as systems embedded in larger systems on which they depend.

The interactions between musicians were the most vivid example of the relational aspect of creativity for me personally because of my own experience as saxophone player. As a saxophone player I depend on any number of companies to produce my equipment: saxophones, mouthpieces, reeds, straps, and other essentials. As a saxophone player I am part of a musical tradition of instrumentalists dating back to Coleman Hawkins, Ben Webster, Lester Young, up to the present. I am part of larger musical ecosystems that include jazz, funk, rock, western classical music and reggae, each with their own musical vocabularies, traditions, and iconic players. I am dependent on the recordings of these great players and composers, most of whom I would not have heard if it hadn't been for the development of recording technology, record companies, distributors and that earthly paradise lost for the musician young and old, record stores. I am dependent on music venues to hear live music and to perform, on teachers and on musicians who particularly inspire me. Saxophone performances almost always involve another player to provide chords: a guitar or keyboard player, if not a whole band. Solo saxophone performances and recordings are few and far between and seem to occur mainly on city street corners. The dependence on chords applies to the trumpet, the oboe, the bass, and most other non-chordal instruments. Playing the saxophone is therefore a relational experience that occurs in a network, a larger musical ecosystem, if you will. A piano or guitar player on the other hand can play chords and be relatively self-sufficient in performance. Nevertheless Purser and I argued that even solo performance remains fundamentally relational, not least because it includes an audience, and missing an audience, the musical network in time and space is always in the musician's head, heart, and hands.

The point of our article was not to deny the existence of genius, of individuals of extraordinary ability. It did question the lone aspect, the suggestion of creation ex nihilo, and the impression it gave that ultimately other people, history, and the environment didn't matter. Tony Kushner, "author" of *Angels in America* wrote an essay asking "Is it a fiction that playwrights write alone?" (Kushner, 1997). Kushner certainly acknowledged he was the author of the play, but an author in a larger web or network of relationships without whom the work could not have emerged. Purser and I felt strongly that the almost exclusive focus on genius had actively prevented an understanding of many systemic aspects of creativity both in popular culture and in academic research because it made an exclusive claim on the *who* of creativity. It also created a certain abstract image of the creator that didn't address the realities of creation, or, as creativity research Frank Barron proposed, that all creation is collaboration (Barron, 1999). It was as if the mystery, grandeur, and often the tragedy of genius had blinded researchers to other forms of creativity. It abstracted the person from their environment, and as a result gave a limited and limiting view of creativity.

The response to our article was immediate. Criticisms showed up in the very same issue, starting with the journal editor's commentary (Greening, 1995), and an article on literary genius (Hale, 1995). Both argued that we were for the social and against the individual. Both were essentially calling us sociological determinists. At a conference soon after we were called socialists and even communists. Apparently we had touched a nerve, a nerve that triggered polarization.

The criticisms we faced were very instructive because it became clear that the raw nerve we had touched was part of a larger and somewhat frazzled nervous system, a network or constellation of interconnected ideas that militated against our contextual, systemic perspective and formed a core of American cultural identity. This core included a number of binary opposition such as individualism versus collectivism, analytical versus holistic thinking, atomism versus holism, methodological individualism versus methodological holism, and capitalism versus communism or socialism (Laszlo, 1996; Montuori & Purser, 1999a; Nisbett, 2003; Phillips, 1976; Stewart & Bennett, 1991; Triandis, 1995). Through that lens, when Purser and I critiqued what we felt was the overemphasis on the individual, we were by definition collectivists, sociological determinists, and so on, because for our critics there seemed to be only two alternatives. Either you're for the individual, or you're for the social. In this article, for the sake of convenience I'll refer to this the constellation of oppositions as being between individualism and collectivism.

Individualism and the End of Man

The American philosopher Fredric Jameson (Jameson, 1983) describes two interpretations of the fate of the individual in postmodernity. The first argues that there may have been such a thing as individualism, at the beginning of capitalism, with the formation of the nuclear family and bourgeoisie, but today that individual does not exist anymore. In the second, more radical view, individualism in the form of autonomous, independent subjects never existed. In fact, individualism is a ruse that gives people the feeling they are free individuals, but really makes them completely dependent on economic and political powers that make them feel good with the rhetoric and flag-waving of freedom and the opportunity to give their hard-earned money to corporate interests.

Regarding Jameson's first position, we are reminded that in the 1950s works such as Whyte's *The Organization Man* and Riesman's *The Lonely Crowd* had already proposed that Americans were a lot more conformist than was previously thought (Riesman, 1950; Whyte, 2002). In 1970 Philip Slater's classic *The Pursuit of Loneliness* also highlighted the conformism of American individualism. As a result of this individualism and the quest for independence, Slater found that Americans had become disconnected, bored, unprotected, unnecessary, and unsafe (Slater, 1990). Mary Catherine Bateson would have approved of Slater's summary: "Individualism is rooted in the attempt to deny the reality of human interdependence," to which he added that "human beings are interdependent, they can only *pretend* not to be" (Slater, 1990, p. 30).

While its philosophical origins were different, Jameson's second position is not actually that far removed from Mary Catherine Bateson's. It asserts the impossibility of a certain kind of autonomous individual, untethered to the point of being unmoored. It also adds the view that a certain kind of (hyper-)individualism is actually an ideology that convinces people they are free: free to choose, free to be who they want to be, free not to wear masks during a pandemic, free to buy products, free to follow trends, free to be like everybody else. In fact, the myth of individualism is being used to manipulate those very individuals for political and economic reasons by those in power. The illusion of independence and consumerist individualism as the embodiment of the American Dream has become the opium of the American masses. "Don't tax the rich," a San Francisco homeless person told my colleague Michelle Marzullo recently, and very seriously and without a hint of irony added, "because that could be me one day!"

Reframing the Discussion

The philosopher Jay Ogilvy (Ogilvy, 1992, p. 229) has argued that "rather than seeing the individual and the collective as ontologically given and concrete, individuality and collectivity can be recast as equal and opposite abstractions from the concrete life of everyday communities."

This is an important first step. Ogilvy helps us by breaking open two closed, homogeneous systems, individualism/collectivism, moving beyond ideology and opposition to the reality of everyday life. What is particularly important here is that the illusion of ontological concreteness can also prevent a questioning of actual experience—it establishes as a done deal a thin interpretation of an experience that is in fact enormously complex.

What becomes apparent in reviewing the literature (Oyserman, Coon, & Kemmelmeier, 2002) is that the entire individualism/collectivism construct originates from a traditionally Western, atomistic/individualistic perspective and portrays collectivism as a rather undifferentiated, homogeneous whole in opposition to the individual. Not surprisingly, the construct of collectivism has been criticized most sharply by Asian scholars (Kim, Lim, Dindia, & Burrell, 2010). An argument has been made, for instance, that rather than individualism/collectivism, a difference in *worldviews* is the key issue. A holistic rather than analytic worldview better represents the cultural foundation of East Asian cultures (Lim, Kim & Kim, 2011). Nisbett and Miyamoto summarize the key differences between analytic and holistic views:

> Westerners tend to engage in context-independent and analytic perceptual processes by focusing on a salient object independently of its context, whereas Asians tend to engage in context dependent and holistic perceptual processes by attending to the relationship between the object and the context in which the object is located (Nisbett & Miyamoto, 2005, p. 467).

Extensive research on individualism and collectivism has shown that a simple opposition does not reflect the complexity of the relationship between the individual and the collective: A meta-analysis shows there is an increasing breakdown in the sharp division between individualism and collectivism (Oyserman et al., 2002). It shows, for instance, European Americans to be less individualistic than African Americans or Latinx, that "Americans are relational and feel close to group members, seeking their advice" (p.72). Brewer and Chen summarize it thus:

> Whereas people in Western individualistic cultures tend to place emphasis on the categorical distinction between ingroups and outgroups, people in East Asian cultures tend to perceive groups as primarily relationship-based. … It is primarily Western European and North American individualistic cultures that rely heavily on abstract, categorical group memberships in constructing social identities. For people in East Asian collective cultures, the primary source for identification and cooperation emanates from the maintenance of relational harmony and promotion of cohesion within groups. Accordingly, it is theoretically important to differentiate between relational collectivism and group collectivism (Brewer & Roccas, 2001) as two different forms of social embeddedness (Brewer & Chen, 2007, p. 137).

Drawing on Brewer and Chen's work, Lim et al. describe the two types of collectivism:

> Group-based collectivism stresses obligations to a group as a whole, valuing obedience to group norms and authority, and subordinating individual interests to those of the collective. Relational collectivism, on the other hand, emphasizes relationships, mutual cooperation, dependence, and concern for each other in a closely interconnected social network. (Lim et al., 2011, p. 24)

Group-based collectivism, they write, is more common in the United States, whereas relational collectivism is found more frequently in East Asia. Interestingly, group-based collectivism sees the group as a source of authority to be obeyed, which might explain Americans' ambivalence towards groups and the collective, while relational collectivism seems to be more relational and focus on mutual aid. The relational collectivism discussed by Brewer and Chen is in fact a much more nuanced network of relationships rather than a homogenizing force.

At the level of everyday life Ogilvy speaks of, this suggests there is a spectrum of human possibilities between individualism and collectivism. We might say that moving along the spectrum of possible behaviors can be activated as a function of culture, social context or by choice. In other words, individualism and collectivism are not static, fixed abstractions, but dynamic patterns of human possibilities.

Ogilvy argued for thinking that goes beyond all or none and instead lands on *some*. He continued his reflection on individualism and collectivism suggesting that one place to find this some is in groups: "Starting with the concrete life of limited groups, it should be immediately apparent that the group needs to optimize the strengths of both individualism and collectivism, while at the same time minimizing the dangers of each extreme" (Ogilvy, 1992, p. 232).

Using Ogilvy's suggestion as a starting point we can see how a classic jazz group can serve as a particularly useful example of such a group (Montuori, 1996). The individualism/collectivism duality sets up an either/or choice, where either the individual or the collective is central and privileged. We can see the pervasiveness of this kind of thinking in a popular slogan found in American businesses eager to promote the importance of teamwork: "There is no I in group." The slogan tells us is that in order to do good teamwork we must completely submerge our individuality in the mission of the group. This sounds very much like group collectivism. Jazz is one context that provides an excellent example of why this form of participation in groups need not be the only choice.

In a classic small jazz group, the individual is not completely submerged in the group (Berliner, 1994). The goal is not to have faceless musicians. In fact, the individuality of the single players is particularly valued, because an individual's unique sound contributes to the overall complexity and richness of the sound of the group. As an example, there is the classic Miles Davis quintet of the late 1950s that recorded *Kind of Blue*, one of the most important as well as bestselling recordings in the history of jazz (Kahn, 2007). The band was composed of players who all had strong musical personalities. Miles Davis, the leader of the session, was already established as a major figure in jazz. Saxophone players John Coltrane and Cannonball Adderley were players with strong individual voices and masters of their instruments who would soon become renowned leaders. Drummer Jimmy Cobb, bassist Paul Chambers and pianist Wynton Kelly formed a well-established rhythm section, and Bill Evans, later to be recognized as a great innovator, took over the piano for several numbers. The larger point is that whether the individual members of the group are famous, soon-to-be-famous, innovators, or not particularly innovative or famous at all, the art form of jazz requires both the ability to perform and solo with one's individual voice and to be able to support the band as a whole by playing in a way that contributes to the whole coherently and reflects the overall sound of the band, while still doing this with one's own unique musical sound.

There is a cybernetic process of navigating extremes whereby the skilled player avoids overplaying, grandstanding, and drawing all the attention to themselves, but also avoids just phoning it in—playing in a way that is rote and full of clichés. A great rhythm section player—a piano player, for instance—doesn't just get to shine during their solo, but also in their ability to *comp*, to inspire and make the soloist sound good (Berliner, 1994). This is something that may not be obvious to the untrained listener whose attention is likely focused on the soloist, but among musicians the ability to support other musicians is highly prized. While comping the piano player is still doing so with their unique sound and choices of chords, use of space, and so on. The complexity of relationships, of interconnectedness, interdependence, and creativity in a creative group is considerable. It goes way beyond simplistic dualisms and is an area where much more research is needed (Paulus & Nijstad, 2003; Sawyer, 2007).

There is tension in the life of any group, whether a jazz band, a home owner's association meeting in a small apartment building, a terrorist cell, or the board meeting of a corporation. Is an individual taking up too much time? Pushing their own agenda too much? Talking too long? Are there freeloaders who don't contribute and leave others to do the heavy lifting? Do the group members listen to each other? Are some individuals having to compromise too much, or not at all? Whose interests are not being served? There are paradoxes of group life, and one of these is the paradox of identity (Smith & Berg, 1987), or the struggle between the group identity and individual identity.

Creative groups show the possibility of individuals who are both asserting their individuality and integrating into a group. It's a question of some rather than all or nothing, individuality or submission/submersion. In the *Kind of Blue* Miles Davis band, the individual members of the group benefited from performing in a band that made everyone sound better and allowed individuals to shine but also reflected Miles Davis's singular vision. A band can be more than the sum of its parts, as was the case with *Kind of Blue*, but also less than the sum of its parts (Morin, 2008b), when the whole inhibits the individuals too much or the individuals are too focused on themselves and not enough on the sound of the band as a whole. Creative groups have much to offer in terms of understanding human possibilities as well as human limitations. Along with some rather than all or nothing, they also teach us about the importance of *sometime* rather than always or never. When is the right time to step forward or move back?

Introducing the dynamic process of creativity can assist in moving beyond the individualism/collectivism opposition, not least because it provides examples of how it is possible to create a mutually beneficial relationship between the musicians, where the individual serves the group and the group serves the individual. This means there is not a static, either/or relationship between individual and collective, but a process over time that, involves periods of integration and periods of differentiation in the context of a network of individuals. Optimal distinctiveness theory (Leonardelli, Pickett, & Brewer, 2010) posits that human beings have two contrasting needs, for assimilation and inclusion into a group and also for differentiation. The two opposing motives can produce an emergent characteristic, namely the capacity for social identification with distinctive groups that satisfies both needs simultaneously. In the case of jazz this is particularly apt because although historically there have been working bands that stay together for extended periods of time, it is more usual for musicians to draw on a network of players for recordings, individual concerts, and tours. Networks extend way beyond geographical proximity as recordings can be made by musicians who are never in the same room. Current file transfer and recording technology makes it easy for a musician in one country to receive a track by electronic file transfer, record one or more takes of their part, and send it back to the artist(s) somewhere else on the planet, something which is quite common these days, and has been a lifeline for many musicians during the pandemic.

Networks, Reinvention, and New Directions in Individualism

The networked world is changing global culture (Castells, 2009; Christakis & Fowler, 2009). What is happening to young people who have grown up in a networked world? Two books about Millennials in the United States give dramatically different diagnoses: *GenerationMe*, subtitled *Why Today's Young Americans Are More Confident, Assertive, Entitled—and More Miserable Than Ever Before* (Twenge, 2006), And *Generation We,* Subtitled *How Millennial Youth Are Taking Over America and Changing Our World Forever* (Greenberg & Weber, 2008*).*

GenerationMe focuses on what the author, psychologist Jean Twenge, finds to be the problematic aspects of hyper-individualistic young people today. Millennials are narcissistic, entitled, they have unwarranted confidence in their own abilities, and their helicopter parents are to blame. In *Generation We*, the authors find the same Millennials are community oriented, dedicated to social causes and driven to make changes and sacrifices during the planet's hour of need. So what are these Millennials, besides the fact that they now have Gen Z snapping at their heels? Selfish individualists? Altruistic collectivists? 21st century schizoid Gen?

Part of the apparent contradiction is due to the fact that these assessments may both be correct. In other words, Millennials and Generation Z in particular appear to be both more individualistic and also more collectivistic or community-oriented than their generational predecessors. Gen Z have been found to be the most individualistic generation yet, but 70% of Gen Z who were polled also agreed with the statement "The wellbeing of society is much more important than my needs as an individual" (Howard, 2018). If they were not also the most individualistic generation in history, one would think Mao finally got through to them and Ayn Rand is rolling over in her grave. Gen Z also appear to be connected to multiple niches rather than one clique, a form of *networked individualism*. Individuals are then nodes in a network of relationships and niches, rather than members of one specific group. This largely because the internet has given them access to a much wider range of relationships, potentially all over the planet.

New forms of individualism are emerging, side-stepping the opposition of either individualism or collectivism. Rainie and Wellman, who coined the term networked individualism (Rainie & Wellman, 2012), emphasize it requires new ways of thinking. They point out that networked individualism was missed by academics because they saw the world in an either/or way, and the assumption was that people either related in group or functioned as individuals. They added that theirs is part of a movement against the Aristotelean view that the world can be structured in groups which in turn can be classified and subdivided with Linnean neatness.

The networked self reflects the emergence of a networked society. Whereas in the past groups were largely based on geographical proximity, on neighbors, neighborhoods, occupations and schools, thanks to technology the new, multiple loose knit niches can be distributed all over the world. The web has been instrumental in creating a great compression of time and space, suggested by Harvey in his pre-internet *The Condition of Postmodernity* (Harvey, 1991). Along with the greater interconnectedness, everything happens faster and appears closer. But interconnectedness is not enough. As the physicist Paul Davies reminds the new paradigm is both systemic, addressing the interconnectedness of the universe, and creative.

The sociologist Anthony Elliott has also proposed a new individualism with "four core dimensions: a relentless emphasis on self-reinvention; an endless hunger for instant change; a fascination with social acceleration, speed and dynamism; and a preoccupation with short-termism and episodicity" (Elliott, 2013, pp. 49–50).

The phenomenon of self-reinvention is key here, and the other three core dimensions seem to spring directly from the compression of time and space. It is a world of instant communication, where everything is moving faster, where the short term rules, where change is desired and expected to be immediate. It's no surprise that the self too must change, and individuals are reinventing themselves. They are reinventing everything from their bodies to their careers, and we should keep in mind here that in a rapidly changing economy, career reinvention is often a stark necessity. Women and men reinvent themselves through workshops, self-help manuals, meditation, therapy, coaching, yoga, dieting, cosmetic surgery, shamanic journeying, psychedelic explorations, and a host of other transformative activities. Embedded in multiple networks, exposed to an overwhelming amount of information, multiple cultures, sub-cultures, and trends, individuals have almost endless choices as long as they have the cash to spare, don't mind that not everything sticks, and that the cosmetic surgery can make everybody look the same.

The point is that along with this networked self, there is also a continually reinventing self, a different self for every niche. Today's choices are dizzying, but what are the criteria for reinvention? What constitutes an improvement? What is better? Is different enough? Is it more Likes? How is human creativity being channeled? Already in 1976, Philip Slater (Slater, 1990) remarked about Americans that "every morning all 200 million of us get out of bed and put a lot of energy into creating and re-creating the social calamities that oppress, infuriate, and exhaust us" (p.2). Perhaps the emerging question is: What are human beings creating? Can human beings take responsibility for their ongoing creation, reflect on it, and make wiser choices?

We recall Ogilvy's remark that "The pressure toward postmodernism is building from our lack of ability to overcome certain dualisms that are built into modern ways of knowing" (Ogilvy, 1989, p. 9).

Since those words were written, postmodernism's popularity has taken a bit of a nosedive, and nobody has much of an idea of where we're all headed although the general sense seems to be that it's nowhere particularly good (Berman, 2011; Greer, 2016). Ziauddin Sardar's (2010, 2015) argument that we're in post-normal times, faced with chaos, complexity, and contradictions, seems most appropriate. The dualisms and contradictions are still there, with individualism and collectivism reaching a fever pitch of ideological polarization, at least in the US. At the heart of the issue for a good portion of individualists is the extent to which the government imposes regulations on them, telling them what they can and cannot do. And yet some groups that celebrate freedom and individualism insist on the government determining what a woman does with her body. Chaos, complexity and contradictions indeed. Whose freedom? Group collectivism again views the group, government, or more generally the collective as an imposed authority on the individual. In Japan's more holistic, relational collectivism, mask-wearing was common before the pandemic, and it is not compulsory during the pandemic. Masks continue to be worn because the assumption is that they will protect others (Nakayachi, Ozaki, Shibata, & Yokoi, 2020). The Xi government has started a battle against inappropriately successful capitalism and manifestations of individualism in newly mediated China (Ni & Davidson, 2021), while in China there is a marked shift towards the dreaded individualism (Sun & Ryder, 2016).

The remarks from Mao and Smith earlier point to the complexity of the phenomena of individualism and collectivism. The constellations I have referred to are not purely cultural (individualism) or cognitive, such as methodological individualism or simply seeing the individual as the unit of analysis but there is an added political/ideological dimension, which has typically been ignored in much of the cross-cultural research and plays a key role. The quotation from Mao made it very obvious what was to be expected from a good communist citizen. Being a bad citizen was not an option. Sacrificing the self for the collective was required in the authoritarian context of Mao's China. In the case of individualism, Jameson reminded us of the way individualism was a philosophical and cultural mystification that deluded people into believing they were autonomous subjects but was in fact a way to manipulate them as political and economic subjects. The result is a culture and a sense of self based on lack (Loy, 2012), a culture of consumption and of never-enough where one's individual worth is based on how much of oneself one has paid into the economy and whether one is keeping up with the Joneses (Stewart, 2021). It behooves us also to remember that the core role models of American individualism in popular culture have historically all been white men. The quest for un-relational freedom surely applies to them, but much less so to women and people of color.

Beyond Individualism and Collectivism

As we move towards the concluding section of this paper, I would like to step back and return to the larger picture. The networked society has created a greater awareness of interconnectedness, but what are human beings doing with this interconnectedness? We have learned that the whole oppositional, dualistic, polarized edifice of individualism vs. collectivism is crumbling. Individualism and collectivism are not simple, monolithic, static, homogeneous, constructs. Individualism comes in various shapes and sizes, with various degrees of selfishness, consumerism, and conformity. Even individualists have elements of collectivism. Individualist America has a healthy dose of conformism. Individualism itself has arguably become an ideology to conform to. The mythology of individualism is fading, and it turns out that the lone cowboy, the lone private investigator, and the self-made man where, like the creative genius, few and far between. What of the rest of the population? They have been dreaming the life, but not living the dream. Rather than lone heroes they have become collective consumers.

The concept of collectivism was built on shaky ground, with "the collective" in the research literature actually referring mostly to networks of interpersonal relationships (Brewer & Chen, 2007). Constructed by Western minds in opposition to individualism, it was articulated with not enough nuance to accommodate that it also comes in various forms, and with its own individualistic tendencies. In the United States, the group and/or collective are viewed as authorities imposing norms on the individual, and therefore often viewed with suspicion. This is one reason why examples of relational creativity can assist in showing other possibilities and serve as a attractor images, showing what is possible.

Brewer (Brewer, 2004) has been clear that psychologists have not adequately studied the implications of interdependence, and should remember that it all the building blocks of human experience are shaped by social interdependence. Her argument takes us back to Mary Catherine Bateson, and the opening salvo from physicist Paul Davies about the new paradigm in science, a systemic and creative paradigm.

Let us assume with Mary Catherine Bateson that independence is a myth, that human beings are interdependent, and humanity's relationship with the natural world is likewise one of interdependence. Let us assume, furthermore, that because of the networked society, younger generations—Millennials and Generation Z—are growing up in a networked world and with a greater appreciation for interconnectedness and relational creativity. We return to the question, what does that mean in terms of what they choose to do?

The French philosopher Edgar Morin has argued for the importance of a planetary ethic, and stated that

> a planetary ethic is an ethic of the concrete universal. The fragments of humanity are henceforth interdependent, but interdependence does not create solidarity; they are in communication with one another, but technological or trade-based communications do not necessarily create understanding; the accumulation of information does not necessarily create knowledge, and the accumulation of knowledge does not necessarily create understanding. (Morin, 2004, p. 24)

In other words, there is nothing about interconnectedness or even interdependence that necessarily guarantees *solidarity*, which we might define as Morin's term for the embodied experience of interdependence. To go beyond the myths of individualism and collectivism requires an anthropological reflection that clearly illustrates how these two terms are both past their sell-by date, begins to show alternatives to the dead-ends that this dualism leads to, and suggests new ways of being, relating, knowing, and doing. There is a need for new ways of making sense of existence and experience, drawing, on the concrete experience of human beings and a sense of responsibility for who and what human beings are creating, an why (Washington, 2003). This is a key challenge for human creativity and the development of a planetary ethic.

Academia and popular culture were blind to the reality of creative groups and relational creativity. Now a more relational creativity is increasingly in the spotlight, through new approaches in psychology and its relevance and popularity in the business world (Anderson, Potočnik, & Zhou, 2014; Bennis, 1998; Glăveanu, 2014; Trompenaars & Hampden-Turner, 2010). The experience of relational creativity can be used to illustrate a way beyond the impasse of individualism and collectivism. It can show how it is possible to remain unique and individual and creative in the context of groups and communities—that it is possible to go beyond either/or. On closer reflection this may not seem like a particularly radical suggestion. Nevertheless an ideological and cultural curtain blinded Americans in particular in the same way when relational, social creativity was not seen (Montuori & Purser, 1999b). Creativity must also be conceived more broadly in the context of the need to create different ways human beings relate to each other. Relational creativity therefore also refers to creating new ways of relating that are mutually beneficial.

New systems-cybernetic approaches to creativity stress the multidimensional, open system, complex nature of relational creativity (Córdoba-Pachón, 2018; Montuori, 2011b; Trompenaars & Hampden-Turner, 2010), but the challenge for these approaches is to dive into the multi-disciplinary complexity of putting flesh on the bones of such a view, re-connecting the sciences and humanities, and beginning to articulate a new way of knowing and being human, in dialogue with generations who are networked-natives, so that they can share their experiences and benefit from existing research that can articulate the conceptual foundations of interdependence (Morin, in press). The Buddhist systems scholar Joanna Macy illustrates the similarities between the mutual causality in Buddhism and general system theory (Macy, 1991), but also makes it clear that it is not sufficient to simply read a book to actually experience dependent co-arising and the interdependence of all things. Any transformation of the way human identity is conceptualized has to be grounded in lived experience and a practice to break out of habitual patterns of behaving and thinking.

The challenge of breaking beyond the impasse of individualism and collectivism, of breaking open the entire static constellation into a more nuanced world of possibilities and self-creation, is considerable. There are clearly already changes occurring. The networked natives are arguably the vanguard of what Inglehart called the silent revolution in values, a transformation humanity's relationship to the environment, gender, and race and ethnicity (Inglehart, 1977, 2018). Systems theoretical and cybernetic scholars can make a vital contribution by providing the concepts and the language necessary to understand the condition of interdependence and to explore their existential implications for human co-existence.

Conclusion

I started with a quotation about the new paradigm in science, and also referred to the current time of transition, the post-normal era when one world is dying, and a new one has yet to emerge. How is this transition to be approached? How can new ways of knowing and relating help to create this new world? How can the old, outdated ways be left behind? Mary Catherine Bateson writes, "to get outside of the imprisoning framework of assumptions learned within a single tradition, habits of attention and interpretation need to be stretched and pulled and folded back upon themselves, life lived along a Möbius strip" (Bateson, 1994, p. 43).

Inevitably, Mary Catherine reminds us, it is necessary to learn the way the assumptions of a given tradition, whether a culture or an era can blind one to other possibilities. Reified or ontologized they become just the way things are, without an awareness that they too are human creations, and as a result the possibility of alternatives is not considered. We saw an example of this with the fixation on genius, which in turn was related to what I have called a constellation of beliefs that include individualism, analysis, and more.

As deeply rooted as the assumptions are, Mary Catherine offers hope:

> Much of modern life is organized to avoid the awareness of the fine threads of novelty connecting learned behaviors with spontaneity. We are largely unaware of speaking, as we all do, sentences never spoken before, unaware of choreographing the acts of dressing and sitting and entering a room as depictions of self, of resculpting memory into an appropriate past. (Bateson, 1994, p. 6)

In the same way that the myth of the lone genius prevented seeing what was always there, and what in fact people were very likely engaged in, namely collaborative forms of creativity, there are other aspects of life—fine threads of novelty—Mary Catherine suggests people are not aware of. Improvisation is often thought of as a mysterious process, but as she points out human beings are constantly improvising conversations and engaging in improvised activities, unaware that they are actually improvising, somewhat like the bourgeois gentleman in Moliére's story of the same name who was pleased with himself because he realized had been speaking prose all his life.

> Men and women confronting change are never fully prepared for the demands of the moment, but they are strengthened to meet uncertainty if they can claim a history of improvisation and a habit of reflection. Learning to savor the vertigo of doing without answers or making shift and making do with fragmentary ones opens up the pleasures of recognizing and playing with pattern, finding coherence within complexity, sharing within multiplicity. (Bateson, 1994, p. 6)

With her penetrating insights Mary Catherine sees that an uncertain, changing future requires reflection and improvisation. Reflection about the way humanity understands itself, how its ideas and beliefs are both openings and cages, about what constitute better futures, and what humanity can hope to achieve. The etymological root of improvisation is the Latin *improvisus* or unforeseen, and points to the ability to not only to deal with the unforeseen but also to create the new, the surprising, the unforeseen, what cannot currently cannot even be imagined. Mary Catherine Bateson has left us much food for thought, and given us precious guidance for creating a future of interdependence and creative collaboration.

References

Anderson, N., Potočnik, K., & Zhou, J. (2014). Innovation and creativity in organizations: A state-of-the-science review, prospective commentary, and guiding framework. *Journal of Management, 40*(5), 1297–1333.

Barron, F. (1995). *No rootless flower: An ecology of creativity.* Hampton Press.

Barron, F. (1999). All creation is a collaboration. In A. Montuori & R. Purser (Eds.), *Social Creativity* (Vol. 1, pp. 49–60). Hampton.

Bateson, G. (1972). *Steps to an ecology of mind.* Bantam.

Bateson, M. C. (1994). *Peripheral visions. Learning along the way.* Harper Collins.

Bateson, M. C. (2015). Norbert Wiener: Odd man ahead. *IEEE Technology and Society Magazine* (September), 35–36.

Bateson, M. C. (2016). The myths of independence and competition. *Systems Research and Behavioral Science, 33*(5), 674–677.

Bellah, R., Madsen, R., Sullivan, W. M., Swidler, A., & Tipton, S. M. (1985). *Habits of the heart.* University of California Press.

Bennis, W. B., P.W. (1998). *Organizing genius: The secrets of creative collaboration.* Perseus.

Berliner, P. F. (1994). *Thinking in jazz: The infinite art of improvisation.* University of Chicago Press.

Berman, M. (2011). *Dark ages America: The final phase of empire.* WW Norton & Company.

Boorstin, D. J. (1965). *The Americans: The national experience.* Random House.

Brewer, M. B. (2004). Taking the social origins of human nature seriously: Toward a more imperialist social psychology. *Personality and Social Psychology Review, 8*(2), 107–113.

Brewer, M. B., & Chen, Y.-R. (2007). Where (who) are collectives in collectivism? Toward conceptual clarification of individualism and collectivism. *Psychological review, 114*(1), 133–151.

Capra, F., & Luisi, P. L. (2014). *The systems view of life: A unifying vision.* Cambridge University Press.

Castells, M. (2009). *The rise of the network society.* Wiley-Blackwell.

Chaharbaghi, K., & Cripps, S. (2007). Collective creativity: Wisdom or oxymoron? *Journal of European Industrial Training, 31*(8), 626–638.

Christakis, N. A., & Fowler, J. H. (2009). *Connected: The surprising power of our social networks and how they shape our lives.* Little, Brown and Co.

Córdoba-Pachón, J.-R. (2018). *Managing creativity: A systems thinking journey.* Routledge.

Davies, P. (1989). *The cosmic blueprint. New discoveries in nature's creative ability to order the Universe.* Simon and Schuster.

Elliott, A. (2013). *Reinvention.* Routledge.

Glăveanu, V. P. (2014). *Distributed creativity: Thinking outside the box of the creative individual.* Springer.

Greenberg, E., & Weber, K. (2008). *Generation We: How Millennial youth are taking over America and changing our world forever.* Pachatusan.

Greening, T. (1995). Commentary. *Journal of Humanistic Psychology, 35*(3), 3–6.

Greer, J. M. (2016). *Dark age America: Climate change, cultural collapse, and the hard future ahead.* New Society Publisher.

Hale, C. (1995). Psychological characteristics of literary genius. *Journal of Humanistic Psychology, 35*(3), 113–134.

Harvey, D. (1991). *The condition of postmodernity: An enquiry into the origins of cultural change.* Wiley-Blackwell.

Howard, K. (2018, December). Gen Z and the challenges of the most individualistic generation yet. *Illume Stories.* Retrieved November 19, 2021 from https://www.illumestories.com/2018/12/gen-z-and-the-challenges-of-the-most-individualistic-generation-yet/

Inglehart, R. (1977). *The silent revolution: Changing values and political styles among Western publics.* Princeton University Press.

Inglehart, R. (2018). *Cultural evolution: People's motivations are changing, and reshaping the world.* Cambridge University Press.

Jameson, F. (1983). Postmodernism and consumer society. In H. Foster (Ed.), *The anti-aesthetic: Essays on postmodern culture* (pp. 111–125). Bay Press.

Janis, I. L. (1983). *Groupthink: Psychological studies of policy decisions and fiascoes* (2nd Edition). Houghton Mifflin.

Kahn, A. (2007). *Kind of blue.* Da Capo Press. (paperback edition)

Kim, J., Lim, T.-S., Dindia, K., & Burrell, N. (2010). Reframing the cultural differences between the East and the West. *Communication Studies, 61*(5), 543–566.

Kushner, T. (1997). Is it a fiction that playwrights write alone? In F. Barron, A. Montuori, & A. Barron (Eds.), *Creators on creating. Awakening and cultivating the imaginative mind.* (pp. 145–149). Putnam.

Laszlo, E. (1996). *The systems view of the world: A holistic vision for our time.* Hampton Press.

Leonardelli, G. J., Pickett, C. L., & Brewer, M. B. (2010). Optimal distinctiveness theory: A framework for social identity, social cognition, and intergroup relations. In *Advances in experimental social psychology, 43*, 63–113.

Lim, T.-S., Kim, S.-Y., & Kim, J. (2011). Holism: A missing link in individualism-collectivism research. *Journal of Intercultural Communication Research, 40*(1), 21–38.

Loy, D. (2012). *A Buddhist history of the west: Studies in lack.* SUNY Press.

Macy, J. (1991). *Mutual causality in Buddhism and general systems theory: The Dharma of natural systems.* State University of New York Press.

McGilchrist, I. (2021). *The matter with things. Our brains, our delusions, and the unmaking of the world.* Perspectiva Press.

Montuori, A. (1996). The art of transformation. Jazz as a metaphor for education. *Holistic Education Review, 9*(4), 57–62.

Montuori, A. (2011a). Beyond postnormal times: The future of creativity and the creativity of the future. *Futures: The Journal of Policy, Planning and Future Studies, 43*(2), 221–227.

Montuori, A. (2011b). Systems approach. In M. Runco & S. Pritzker (Eds.), *The encyclopedia of creativity* (Vol. 2, pp. 414–421). Academic Press.

Montuori, A., & Purser, R. (1995). Deconstructing the lone genius myth: Towards a contextual view of creativity. *Journal of Humanistic Psychology, 35*(3), 69–112.

Montuori, A., & Purser, R. (1999a). Introduction. In A. Montuori & R. Purser (Eds.), *Social creativity* (Vol. 1, pp. 1–45). Hampton Press.

Montuori, A., & Purser, R. (Eds.). (1999b). *Social creativity* (Vol. 1). Hampton Press.

Morin, E. (2004). *La Méthode, tome 6. Ethique* [Method, volume 6. Ethics]. Seuil.

Morin, E. (2008a). *La méthode: Coffret en 2 volumes.* [Method: Boxed set in two volumes]. Seuil.

Morin, E. (2008b). *On complexity.* Hampton Press.

Morin, E. (in press). *The challenge of complexity* (A. Heath-Carpentier, Ed.). Sussex Academic.

Nakayachi, K., Ozaki, T., Shibata, Y., & Yokoi, R. (2020, 2020-August-04). Why Do Japanese People Use Masks Against COVID-19, Even Though Masks Are Unlikely to Offer Protection From Infection? [Brief Research Report]. *Frontiers in Psychology, 11*(1918). Retrieved November 29, 2021 from https://doi.org/10.3389/fpsyg.2020.01918

Ni, V., & Davidson, H. (2021, 10 September). China's cultural crackdown: few areas untouched as Xi reshapes society. *The Guardian.* Retrieved December 10, 2021 from https://www.theguardian.com/world/2021/sep/10/chinas-cultural-crackdown-few-areas-untouched-as-xi-reshapes-society

Nisbett, R. E. (2003). *The geography of thought: How Asians and Westerners think differently … and why.* Simon and Schuster.

Nisbett, R. E., & Miyamoto, Y. (2005). The influence of culture: holistic versus analytic perception. *Trends in cognitive sciences, 9*(10), 467–473.

Ogilvy, J. (1989). This postmodern business. *The Deeper News, 1*(5), 3–23.

Ogilvy, J. (1992). Beyond individualism and collectivism. In J. Ogilvy (Ed.), *Revisioning philosophy* (pp. 217–233). SUNY Press.

Oyserman, D., Coon, H. M., & Kemmelmeier, M. (2002). Rethinking individualism and collectivism: evaluation of theoretical assumptions and meta-analyses. *Psychological bulletin, 128*(1), 3–72.

Pachucki, M. A., Lena, J. C., & Tepper, S. J. (2010). Creativity narratives among college students: Sociability and everyday creativity. *Sociological Quarterly, 51*, 122–149.

Paulus, P. B., & Nijstad, B. A. (Eds.). (2003). *Group creativity: Innovation through collaboration.* Oxford University Press.

Peat, F. D. (2002). *From certainty to uncertainty. The story of science and ideas in the 20th century.* Joseph Henry Press.

Phillips, D. C. (1976). *Holistic thought in social science.* Stanford University Press.

Rainie, L., & Wellman, B. (2012). *Networked. The new social operating system.* The MIT Press.

Richards, R. (Ed.). (2007a). *Everyday creativity and new views of human nature: Psychological, social, and spiritual perspectives.* American Psychological Association Press.

Richards, R. (2007b). Everyday creativity: Our hidden potential. In M. Runco & R. Richards (Eds.), *Everyday creativity and new views of human nature* (pp. 25–54). Ablex/Greenwood.

Riesman, D. (1950). *The lonely crowd. A study of the changing American character.* Yale University Press.

Runco, M. (2007). *Creativity. Theories and themes: Research, development, and practice.* Elsevier.

Sardar, Z. (2010). Welcome to postnormal times. *Futures, 42*(5), 435–444.

Sardar, Z. (2015). Postnormal times revisited. *Futures, 67*, 26–39.

Sawyer, R. K. (2007). *Group genius: The creative power of collaboration.* Basic Books.

Shanahan, D. (1992). *Toward a genealogy of individualism.* University of Massachusetts Press.

Slater, P. E. (1990). *The pursuit of loneliness. American culture at the breaking point.* Beacon Press.

Smith, K. K., & Berg, D. N. (1987). A paradoxical conception of group dynamics. *Human Relations, 40*(10), 633–657.

Staw, B. M. (2009). Is group creativity really an oxymoron? Some thoughts on bridging the cohesion–creativity divide. In E. A. Mannix, M. A. Neale, & J. A. Goncalo (Eds.), *Creativity in groups* (pp. 311–323). Emerald Group Publishing Limited.

Stewart, E. (2021). The problem with America's semi-rich. *Vox* (October 21). Retrieved November 29, 2021 from https://www.vox.com/the-goods/22673605/upper-middle-class-meritocracy-matthew-stewart

Stewart, E. C., & Bennett, M. J. (1991). *American cultural patterns*. Intercultural Press.

Sun, J., & Ryder, A. G. (2016). The Chinese experience of rapid modernization: Sociocultural changes, psychological consequences? *Frontiers in Psychology, 7*, 477.

Taylor, C. (1992). *Sources of the self: The meaning of modern identity*. Harvard University Press.

Traber, D. (2007). *Whiteness, otherness and the individualism paradox from huck to punk*. Palgrave MacMillan.

Triandis, H. (1995). *Individualism & collectivism*. Westview Press.

Trompenaars, A., & Hampden-Turner, C. (2010). *Riding the waves of innovation: Harness the power of global culture to drive creativity and growth*. McGraw-Hill.

Twenge, J. M. (2006). *GenerationMe: Why today's young Americans are more confident, assertive, entitled—and more miserable than ever before*. Free Press.

Washington, J. M. (Ed.). (2003). *A testament of hope: The essential writings and speeches of Martin Luther King*. HarperOne.

Watts, A. (1989). *The Book: On the taboo against knowing who you are*. Vintage.

Whyte, W. (2002). *The organization man*. University of Pennsylvania Press.

Lorusso, Mick. (2009). *Spiral*. Urban Cosmos Series, Meditative Interventions.
Found object photograph. 25 x 30 cm.

Lorusso, Mick. (2006-08). *It Falls Through Me*. Anima Mundi Series, Energy Patterns. Drawing. Graphite on paper. 30 x 22 cm.

Cybernetics and Human Knowing. Vol. 28 (2021), nos. 3-4, pp. 69–83

Chasing the Mind and Body of Metalogue
Catching Recursive Frame Analysis

Hillary Keeney[1] and Bradford Keeney[2]

The literary interaction of the imagined father and daughter characters inspired by Gregory and Mary Catherine Bateson are used to explore recursively structured conversation. We move from the metalogue to our attempts to map and perform change-oriented conversation with recursive frame analysis (RFA). This research method reveals how the analysis of communication is commonly riddled with errors in logical typing, losing touch with or not crossing the gap between action and interpretation. When higher orders of the zig-zag dynamic between form and process are separated or forgotten, the ecological threads of connection are readily broken. We re-emphasize the value of G. Spencer-Brown's ideas on reality construction and in particular the dynamic of recursion, something that gets us closer to the heart of change than a focus on story or narrative alone. In this processual unfolding, the improvisation of speech lines composes the whole circular conversation. Whether the latter is trivial or evocative of transformation has more to do with the whether the performer-analyst is recursively aligned with the Ouroborean loops that are in play.
Keywords: Mary Catherine Bateson, Gregory Bateson, recursive frame analysis, cybernetics, radical constructivism, G. Spencer-Brown

Once upon a time, Mary Catherine Bateson (2003) imagined a conversation with her father entitled, "Daddy Can a Scientist Be Wise?" This exchange, written like a scripted performance, begins with an introductory note:

> My thinking in this essay, written in 1977, reflects the 1968 Wenner-Gren Conference on Conscious Purpose and Human Adaptation, organized by Gregory, about which I wrote *Our Own Metaphor,* as well as later conversations, but I had not yet worked with Gregory on *Mind and Nature.* Here, I explore Gregory's idiosyncratic definitions of evocative terms like "love," "mind," and "wisdom" in terms of a cybernetically-based epistemology. The style and context are reflective of his Father-Daughter "metalogues," composed to explore concepts he was not ready to define fully... (Bateson, 2003, p. 3)

When Gregory Bateson invented the metalogue, he wanted to explore a different conversational means for handling metaphors of cybernetic epistemology. He needed a literary foil and partner to create some dialogical tension so the conversation could generate a difference that evoked ongoing differences. These characters were not meant to battle like symmetrical adversaries, but to introduce a marked distinction within a complementary caring relationship. The counter voice in his metalogues, and in the critical editing of his actual writing, was Mary Catherine Bateson. He valued her because her approach differed from his while they fought on the same side of the

---

1. Benemérita Universidad Autónoma de Puebla and Etfasis Institute of Systemic Family Therapy.
   Email: hillarykeeney@gmail.com
2. Benemérita Universidad Autónoma de Puebla and Etfasis Institute of Systemic Family Therapy.
   Email: bpkeeney@gmail.com

epistemological battlefield. Both daughter and father chased the *muddles* of their time like Don Quixote in search of a cybernetic La Mancha. In this pursuit she sought clarity by making complex phrases from her father's intellectual lineage more readily accessible for the general reader to understand. He sought a different order of precision by extending the breadth of what cannot or should not be understood, often drawing upon cumbersome terms and phrases that were not easily accessible.

In scholarly life, the daughter of the metalogue wanted her father's concepts more clearly defined and expressed, made easier to swallow. This guided how her own metaphors were composed for her well-received popular writing. Gregory, on the other hand, resisted any form readily digestible by the appetites of popular consumption. He, like Alfred North Whitehead, preferred to leave the vast darkness of a subject unobscured for future consideration. Gregory and his daughter were unquestionably on the same cybernetic team that fought epistemological muddles with a circular knight's sword, pen, and eraser in hand. But they differed in how they related to the value of popular clarity versus maverick obscurity.

Gregory most likely abhorred the idea that his most treasured concepts should ever be ready to define fully, yet his daughter pushed for it in each metalogue conversation and edited text. Gregory was passionately frustrated and easily irritated by the way scholars communicated about communication, including himself and his daughter. He regarded human communication and social interaction as unavoidably plagued (and occasionally blessed) with the mishandling and mistyping of abstraction. He was fussy about distinguishing descriptions of simple action from punctuated frames of context, and later in life, he further differentiated and increasingly ambiguated context from its containment in a transcontextual weave. This alteration partly arose as a protest against others counting double binds, a notion he regretted prematurely releasing to the academic world. More critically, he regarded most social science and psychotherapy as the induction of trivial patter with lazy discourse that uncritically conflated description with explanation. For example, the suggestion that crazy ideation is caused by madness which, in turn, is the result of schizophrenia, was something he thought better belonged to a fictitious character of Lewis Carroll. Stacking a tower of names with the names of names—crazy, madness, and schizophrenia—only conjures a fantasy of interpretation that is cognitively hypnotic and epistemologically toxic.

No matter what natural history Gregory Bateson observed, he also listened to how other observers described it in ways that fostered a muddle of misplaced clarity and mistaken context. We sometimes affectionately refer to him as the patron saint of well-typed description. Next to cybernetics, the other main metaphor of his life was likely the term *muddle*. His mission seemed to tease apart the logical typing entanglements or knots of misperception and misconception that removed an observer from discerning cybernetic pattern in variant orders of description and explanation. As an elder in need of a capable partner, he often enlisted Mary Catherine to be by his side in this circular, binocular pursuit of his impossible dream.

Perhaps Gregory created the metalogue so he could invent conversations that probed impossible answers to his unreachable questions. He made no secret that he was inspired by the counsel of e. e. cummings whom he misquoted to make his own point, "always the more beautiful answer who asks the more difficult question" (Bateson, 1979, p. 213). His conversational journey into metalogue was born of improvisation, thematic extension, and the recursive re-entry of altered metaphors that evoked a pattern connecting the participants, including the reader, to greater explanatory uncertainty amidst the wonder and awe of complexity. Whereas he and his daughter had before chased the elusive cybernetic mind of conferences, redwood forests, a blind man walking, and co-evolution, later their metalogues aimed to find a transient interactional mind in conversation. This "mind" was never limited to the body's verbal or nonverbal expression, whether consciously or unconsciously expressed. It was inherent in any cybernetically organized system that made comparable the widest range of transdisciplinary foci, including porpoises acquiring a new trick and families escaping an interactional tic.

In a metalogue, the topic of interest was handled with serious consideration of the cornerstone metaphor originally cast, the logical typing that kept track of different descriptive orders of process and categorization of form, and the zig-zag movements that oscillated between static nouns and dynamic verbs. Both Gregory and Mary Catherine made sure their maps were not confused with the territory while trying to map how their mapping punctuated, accentuated, maintained, or changed the topic as well as its sensory descriptions and theoretical explanations. Again, Gregory pushed for the territory to remain obscure, tempering the risk that his clear circular perception of the whole might breed reductionist misconception by others. Mary Catherine, on the other hand, preferred making obscure ideas more accessible, but she acknowledged that this endeavor risked the reader forgetting that any well-formed reduction was actually a complex production rather than a simple representation. The metalogues of this father and daughter, as Gregory admitted, more often missed than hit the target. But they demonstrated how a parent and offspring nobly spent a career trying to address the gap between map and territory, mind and nature, creatura and pleroma, form and process, informed and performed, rigor and imagination, science and art, conversation and experiential reality construction.

It took another leap past Korzybski for the Bateson duo to avoid mindlessly repeating that "the map is not the territory" and that "linear thought is not circular epistemology" and all the other convenient utterances and clichés to which his work has often been reduced. In the same way that it eventually dawned on cyberneticians to consider that the cybernetician was inside cybernetics, it took a while to appreciate that the cartographer was in the territory and more radically, should also be mapped as the one mapping. Second-order, higher ordering cyberneticians did their best to catch up to the implications of the recursive intertwining of mapper, map-making, and the know-how behind discerning and enacting different orders of map-territory relations. Early on, Gregory Bateson understood that the territory begins in the flux of the ineffable, the Gnostic pleroma that has not yet been transformed by the wordsmithing and knowing of the creature of the mythical Creatura. The vastness of this epistemological/ontological maze was beyond the gaze of common perception and was only revealed when it partially concealed the double binding of minding, knowing, mapping, and conversing. The more the territory is presumed to be caught, the more everyone involved loses their grip as the universe expands, changes, and escapes the measurement pincers, the butterfly net, and every kind of prose that is not the rose.

Gregory Bateson caught the importance of George Spencer-Brown's dynamic of reality construction in the last decade of his life (Keeney, 1979). He saw it as going paradigmatically further than the map-territory differential. Spencer-Brown's calculus of indications made explicit the dynamics of building a reality rather than only posting a warning about the pitfalls of mapping the territory. His contribution replaced the need for the mathematical injunction to use logical typing for the banishment of paradox. However, Gregory and later, Mary Catherine, continued to logically type, though they did so more in the spirit of William Blake—revealing the outlines of form and assessing whether they were an impoverished muddle or contributed to the stochastic processes of higher learning that inspires higher aesthetic yearning.

As scholars and practitioners of change-oriented conversation, whether it takes place in therapy, counseling, social work, diplomacy, community relations, mediation, classroom teaching, cybernetic pontification, higher order attempts of elucidation, or metalogues that seek to soar above ping-pong dialogues, we have been influenced by the contributions of Gregory Bateson and Mary Catherine Bateson and the cyberneticians who invented the ideas this father and daughter applied to everyday phenomenal domains. We continue to be moved by the encircling differences that gave life to their creative interaction in imagined and real time conversation. This is especially the case for our explorations to create and analyze conversations that evoke transformation.

Recursive frame analysis (RFA) is a research method invented for the purpose of tracking the trajectory of change-oriented discourse (Keeney, 1990, 1991; Keeney, Keeney, & Chenail, 2015). Rooted to Spencer-Brown's *Laws of Form* (1969), RFA aims to fulfill the Batesonian quest to reveal logical typing (and mistyping) in human communication, discern cybernetic patterns of interaction, and lay bare the relationship between simple action, description, explanation, context, and transcontextual weave as they unfold in real time conversation. RFA provides the antidote to the mishandling of abstractions and the reduction of cybernetic complexity by choosing to track—rather than try to solve, ignore, or ban—the inevitable dance, stumble, or tumble across the map/territory divide. It tempers the temptation to indulge in excessive abstract explanation or to get lost in handling narratives that come untethered from the action scene. RFA maintains the Batesonian humility that we cannot claim to fully know the ecology of mind as it is embodied and expressed in communication, but also utilizes this fact to prescribe greater freedom to work creatively in the contextual weave of reality construction.

The Fundamentals of Recursive Frame Analysis (RFA)

When a change-oriented conversation begins, the metaphor offered at the start is not yet a frame, context, unfolded story, or enacted performance. It takes time for these larger expanses of sequenced experience to be constructed. We regard the first lines spoken in a change-oriented conversation as only offering a beginning distinction, à la Spencer-Brown, that may set in motion a subsequently elaborated weave of metaphors that eventually constitute a frame and whole conversational reality. In the beginning of a conversation, a metaphor is offered that can serve as a foundational starting point.

For example, a client might propose an opening line like "I have a problem," "I am the problem," "I am unsure if I have a problem," "Others claim I have a problem," "My spouse believes that my problem is the problem," or "I need a solution for my problem." All these variations of an opening utterance distinguish and spotlight the metaphor of problem. The practitioner can then re-distinguish (re-emphasize) the client's metaphor and enter more into its theme, building up further discourse that maintains a focus on a problem-distinguished context. Or another distinction can be offered as a counter metaphor, such as "Those are very interesting shoes," "Your middle name is longer than your first name," or "When was the last time you ate a slice of chocolate cake?" These latter distinctions are likely not to be associated with the client's initial distinction. If these variant distinctions are given further attention and elaboration, we drift away from a problem-distinguished conversational reality and are creatively freed to initiate the construction of an unexpected alternative theme.

The future emergence of a familiar or unfamiliar conversational reality is under construction with the first conversational moves. And whether you feed (re-distinguish) or starve (ignore) a distinction helps determine the contextual direction of the conversation. No way around it—how you participate and interact with presenting metaphors contributes to determining the subsequent reality you and the conversants later step into or out of. The conversational performance stage is initially set with distinctions or startup metaphors that hold the potential for becoming frames. How everyone acts determines what is further distinguished, re-distinguished, extinguished, framed, unframed, and made real or not. We are more responsible than we may have previously assumed for bringing forth mind in the nature of conversational reality. In the beginning of every encounter, a distinction is made and we either accept it or we offer an alternative possibility. Whatever the case, within seconds or minutes a distinction may grow into something larger than a mere distinction—it may move toward becoming a frame that, in turn, marches toward constructing, deconstructing, or reconstructing a whole reality.

A distinction that is re-distinguished becomes more importantly distinguished than before, and this is a nontrivial though often underappreciated outcome. Such subsequent re-distinguishing contributes to a metaphor being experienced as more "real" until it feels concretely noun-like. It is seldom noticed and often forgotten that the metaphor began as a soft abstraction before becoming hardened through social interaction. For example, re-distinguishing (or diagnosing) that a problem is truly a problem leads to the further rigidification of problem-distinguished discourse, whether it examines historical origins, social involvement, attempted solutions, fantasized solutions, cultural (mis)appropriations, narratives, or anything at all related to the problem's absence or presence. As this activity of re-distinguishing or re-indication proliferates, the original distinction moves past being only a distinction. It is growing and on its way to becoming a contextual frame that holds other distinctions and re-indications of a similar kind. In other words, a singular distinction shifts to become a class of related distinctions. In the previous example, a problem distinction is built up through conversation in real time to become a problem frame (Keeney, Keeney, & Chenail, 2015; Keeney & Keeney, 2019).

The problem with problems, whether regarded as internal or external to the agents experiencing them, lies not in their presumed nature, cause, or locale, but in their becoming contextual frames rather than distinctions. Existentially speaking, life hosts the ongoing entry, exit, and re-entry of innumerable problems, suffering, and impoverished experience. As long as life is bigger than its subset of problems, we are able to stir more creative movement with more unexpected, surprising interaction with the problematic parts of our experience. However, when a problem grows into being the primary contextual frame, we find our life unnecessarily constrained by treating a problematic part as if it is a whole. Clients often come to a practitioner in need of a frame reversal where life becomes the primary frame that holds its distinctions, including those that are named problems. Change-oriented conversation aims to reset the container of experience as the whole of life.

RFA is a tool that starts its analysis of change-oriented conversation by marking the primary distinction before it has become a frame for any subsequent distinguishing and re-indication that take place within it. Again, conversations begin with distinctions that can either remain distinctions or, via re-indication, grow into frames. Any frame, once built, can be deconstructed to return to what it previously was—a distinction. Or a frame can be re-contextualized inside a distinction it formerly held or within a new one that arises outside the previous frame of reference. This, in a nutshell, is reality construction in action.

Inspired by G. Spencer-Brown and the Bateson family passion to discern the differences between orders of abstraction, RFA introduces a radically new way of tracking how distinctions, indications (and re-indications), and emergent frames account for the anatomy or architecture of any conversational performance. Here the different orders of distinction that are produced in the conversation are sufficient in and of themselves to account for the built-up realities they construe. No outside narratives, explanatory devices, or interpretive maps are necessary. The latter are simply external discourses that are separate conversations from the conversations they purport to explain. In this sense, explanations, narratives, and interpretations are fantasy discourses that claim to be about another discourse. Like all discourses, they are only self-referentially involved with their own constructions and not necessarily a means of elucidation. There are only distinctions, re-indications, and frames, and as they recursively interact with themselves, they self-reflexively verify their own participation in what they bring forth.

RFA invites research and practice to become two sides of one person who both creates and discerns distinctions, distinguishes indications, and indicates frames. As an uber-radical constructivist orientation, RFA recognizes that the primary form of communication is not description but prescription. More accurately, it catches the recycling of its own form, where the whole of what is caught is the contextual frame of its contents. Paraphrasing Spencer-Brown (1969), an RFA recipe, score, or map—like in cookery and music—reveals the instructions for re-constructing the original conversational reality. An emphasis upon our participation in creating an experiential world orients us to look for the recipes, prescriptions, directions, and injunctions that bring it forth. What RFA offers is a way of distinguishing the essential prescriptions (the distinguishing and re-distinguishing) that called or conjured something into being.

The conversational arts too often degenerate into thinking they are handling maps, narratives, and interpretations without recognizing that doing so is the act of creating that which they assume they are explaining. Action is equally primary to knowing, and unless we remember that it is we who are drawing the distinctions, we blame others for bringing forth that which is assumed to be separate from us. RFA asks us to bring ourselves more fully inside the realities we are working with, and in so doing, take more responsibility for not blaming outside entities—whether genomes, bloodstreams, brains, relationships, family constellations, or cultural narratives—and instead act to serve the differences (via distinctions recursively operating to generate indications that may become frames) that foster transformative growth.

Rather than over-emphasize either the map or territory, utilize the difference between them to inspire, irritate, differentiate, relate, transform, and generate other differences that make a difference in the lived performance. Act inside interaction with others to set in motion the kind of differences that help build an eco-friendly neighborhood mind. Be recursively mindful that indication is the recurrent action of re-making a distinction. Make the shift from description to composition, narration to performance, and interpretation to invention.

We first draw a distinction and if we continue to re-indicate it, two things follow. First, we forget that it is we that drew a distinction and that all subsequent indications are also drawn by us. We easily lose sight of the fact we are always acting to bring a reality forth. Secondly, as we continue re-indicating a distinction, the cascade of indications may condense and give the illusion of being thing-like. The recursive operation of re-entrant distinguishing is the process of reification, giving abstractions an unearned and misplaced concreteness. Admittedly, sometimes the conversion of descriptive abstractions into thingish objects is tactically or practically useful, but at other times it leads us astray.

We easily create a world of suffering if we forget that we are the ones creating the distinctions and frames that comprise our world. Then we believe that wrong interpretations need to be exorcised, attacked, or corrected, and that there exists a liberating narrative, story, interpretation, explanation, or map that can lead us to the promised land. As various wisdom traditions have taught, story-making about life-as-a-story is a source of suffering. A Zen slap in the face aims to empty the fossilized verbiage that clogs the natural flow of the life stream. While human beings tend to think like the structure of a story, they first act or perform to invent the idea of a story.

RFA brings us back to what distinction we threw (rather than knew) in the beginning. The first distinction was cast in a conversation before we became lost in the assumption that any indication is more concrete and primary than the constantly changing performance that keeps on acting in order to distinguish-extinguish what is being distinguished-extinguished. Poetically speaking, a conversation comes to life with the presence of oscillation, vibration, or breath between construction and deconstruction. At the beginning of every conversation, the door to many worlds is open with all participants free to build any reality and at the same time, extinguish other possible realities.

Heinz von Foerster (2003) did not quite complete his aesthetic imperative when he advised that seeing differently requires learning to act differently. More specifically, you must act in order to successively distinguish, indicate, and re-indicate again, until you are caught inside the emergent frame. The experience of a reality does not arise before one's participation in bringing forth a relationship with it. RFA reminds us of how we are forever participating in creation and that we are never outside it as a narrator or map maker who escapes the responsibilities of imposing ourselves onto everything and everyone with which we converse. We both emerge from our conversations and we converge ourselves into conversational space.

Improvising and Composing

Change of any kind in any venue, from the theatre to the therapy clinic, laboratory, or home is rarely evoked through a simple set of instructions. You cannot fully know what to do before you begin. Tinkering, trial-and-error, and the circular self-correction of lower and higher cybernetics get the performance engine moving to a destination that is different than where the journey began. Mary Catherine Bateson regarded this latter way of engagement as improvisation. Our work has been equally inspired by the same metaphor. We formerly innovated "improvisational therapy" (Keeney, 1991) and our subsequent explorations in creative transformation are akin to the art of living life as jazz rather than following static maps, models, and instructions (Keeney & Keeney, 2019). Not only is improvisation the modus operandi for making change happen, it brings life and excitement to the performer. This outlook is summarized by Mary Catherine Bateson:

> Rarely is it possible to study all of the instructions to a game before beginning to play, or to memorize the manual before turning on the computer. The excitement of improvisation lies not only in the risk of being involved but in the new ideas, as heady as the adrenaline of performance, that seems to come from nowhere. (Bateson, 1994, p. 9)

Life is composed through improvisation, that is, the whole composition of one's life performance is achieved by a series of improvised segments and segues. This dynamic applies to the different developmental chapters that constitute a life story, as Mary Catherine Bateson illustrated. Improvisation is also applicable to the composition of conversations that make a difference in steering and turning the course of a life trajectory. On the simplest level, transformative movement occurs when a conversation transitions between a minimum of three distinct contextual frames, corresponding to the three-act structure of a theatrical play: the beginning, middle, and ending.[3] However, there is more to composing conversations and lives than walking in a straight line from "in the beginning" to "they lived happily ever after" having survived a crisis, transition, or leap across the middle. Moving from anywhere to any there involves a circular dynamic that generates a mobilizing difference. A line of conversation connects to other conversational lines by how it bends in other directions, circles back to generate variation, and builds a circular momentum that delivers surprise to both familiar order and unfamiliar disorder.

Arguably, one of the flaws of Mary Catherine Bateson's analysis of personal life composition was her over-emphasis on *story*. Here half the zig-zag ladder that moves back and forth between form and process too often seems missing. The action process that is fodder for a post hoc *story form* is found in the enactment of everyday performance, a dramatological rather than narrative domain. Life is first performed and then told later as a story of what formerly happened. More accurately, daily life is improvised action which includes story-making that uses bits and pieces of former scenarios that are edited to fit each situation. Emphasizing the circular relationship between process and form, or action and story, contrasts with the popular myth that narratives strictly script living.

The history of psychotherapy has vehemently argued back and forth between how to handle the perceived difference between the performing body and the interpreting mind. Brad's work (Keeney & Ross, 1985) originally addressed this split in terms of what we called semantic and political frames, held as a complementarity without preference for one at the exclusion of the other. Bateson's (1979) zig-zag ladder demonstrated how this complementarity may be spelled out for different orders of action description and explanatory abstraction. However, his general aversion to action, especially that meant to evoke change, left him pondering why he rarely embodied the circularity he presumably understood.

Gregory Bateson's natural history emphasis anchored him to observation rather than untethered flights of intervention, as he resisted acting in the service of change. In private conversation with Brad (Keeney, 1976) he explained that this allergy stemmed from shame associated with his participation in military intelligence during

---

3.   While Mary Catherine Bateson did not analyze the conversational structure of therapeutic sessions, she did so for some of the important conferences she attended. In her book, *Our Own Metaphor* (2005), she laid out how the conference became a self-organizing story. She did the same for a conference organized by Gregory Bateson and Bradford Keeney where she was a major participant, concluding that the conference was itself "the unfolding of a story" (Keeney, 1983b).

World War II. It later extended to his repulsion at therapists and social scientists who over-emphasized the unilateral influence of power over the cybernetic interdependent network of mind, and his embarrassment of being granted privileges he thought he didn't deserve due to his family name.[4] Mary Catherine Bateson was more action oriented like her mother, Margaret Mead. Her insistence on clarifying abstract ideas was, at least in part, to rally social action that made a difference in human rights, diverse relations, and planetary health. While her lean was toward emphasizing composing a life story, she still tended to de-emphasize the nuts and bolts of the improvisational performance that constructs whatever story later emerges—favoring the form rather than the dynamics of its construction.

Gregory and Mary Catherine Bateson long advocated that correct action followed correct perception. If one could see the world cybernetically, systemically, or ecologically then less systemic violent action would ensue. This resulted in their advocacy for acquiring wisdom before daring to intervene. Neither Bateson seemed to grasp or accept the tectonic shift offered by Heinz von Foerster who turned the world upside down when he advised that "if you desire to see, then learn how to act." Nothing could be more challenging to Gregory than this imperative. At the same time, his zig-zag ladder left conceptual room for the arrow of causal directionality to alternate back and forth between action-led and perception-led living, each providing correction and alteration in the body and mind of a circular performer and thinker. In this regard, he perhaps went further than the sleight of hand performed by Heinz von Foerster, the cybernetician-magician, but this was only true for Bateson's observation rather than his stage performance. This Batesonian back-and-forth *viewing as multilevel doing* takes place through the dynamic of recursion so that former zigs are part of

---

4.   Gregory Bateson was surprisingly candid about his family relations, including his disclosure in David Lipset's (1982) biography that he never felt he could meet his father's expectations. Nothing left him feeling more uncomfortable than adulation, expectation, or exaggeration due to his relation to the William Bateson name. Later, his next intellectual father, Norbert Wiener, was equally discouraging when he advised Gregory to not apply cybernetics to the social domain of human interaction. Also troubling was the breakup of Norbert Wiener with Bateson's other mentor, Warren McCulloch, due to the latter's Bohemian lifestyle that upset Mrs. Wiener. Gregory's marriage to Margaret Mead held stylistic differences and argumentative interactions, sometimes resourceful, that were also made public in their conversation, "For God's Sake, Margaret" (Brand, Bateson & Mead, 1976). Finally, his relations with his Palo Alto colleagues did not end well relationally or epistemologically, which largely remained a well-kept family secret in the family therapy profession Mary Catherine Bateson grew up in the middle of all these intersecting, complex, and changing family and professional relations.

   Suffice it to say that the context of the Bateson family and their social ecology influenced how they defined, explained, and influenced ideas concerning context, relationship, ecology, epistemological correction, cybernetic calibration, and the metaphors battling between clear representation and perturbing evocation. That is a future subject for historical study. Gregory shared his personal feelings, stories, and concerns about his family and professional relations with Brad, who at the time was a young family therapist and cybernetic enthusiast. He seemed to believe that his relational ecology was an important context of his ideation. We mention a few of these historical details to respectfully acknowledge what existentially held his metalogues, dialogues, monologues, and diatribes. Let us also not forget that Gregory did not exist in a vacuum. Though today his enthusiasts sometimes act like he was the inventor of cybernetics, he only extended the original thinking of his mentors, mostly into the social domain. Mary Catherine Bateson did the same, perhaps with less rigor when it came to sounding the noise needed for epistemological change, but with more vigor when it came to getting the signal through.

subsequent zags and vice versa as the ladder is climbed. A recursive performance is the key to ending the curse of the checkmate debate over whether action or story is the primary driver of experience; they are circularly intertwined.

Through decades of training practitioners and researchers, we have found that observers unfamiliar with implementing change-oriented conversation are not reliably able to identify the kind of conversation that qualifies for the application of RFA. The ability to define a metaphor, frame, or system of communication does not necessarily indicate a capacity to distinguish their construction when analyzing conversation. We propose that attaining the skills of RFA requires learning how to construct (and deconstruct), sustain (and interrupt), and alter (and stabilize) different contextual frames in live conversational interaction. Once again, Gregory Bateson's invention of the metalogue provides an example of his taking action to invent that which he wished to study. His baklava-like layered conversation aimed to embody the dynamics that were isomorphic to the those of the phenomenon being discussed.

Change-oriented conversation, like the scripts of theatre and metalogue, transitions from one frame or action scene to another. When less skillfully constructed, a conversation may only generate a scatter plot of distinctions, indications, and frames that do not build on one another or connect in way that makes any difference other than exercising social greeting, maintaining a difference in social roles, or simply generating mindless chatter with neither clear signal nor meaningful noise. The RFA performer-observer must be able to immediately distinguish between a transformative conversation that transitions through progressive contextual frames with a clear beginning, wobbly middle, and a transformed ending from one that communicationally meanders or goes nowhere. In addition, one must discern how improvised lines connect through circular means of composition, decomposition, and re-composition.

Many conversations never leave the opening act and only explore different content within the same contextual frame. While an RFA study can explicitly reveal whether and how contextual change was constructed and performed, a detailed analysis of a lifeless and change-less session brings little more than a description of a non-transformative conversation. Here there is no trajectory of change to score—the conversation remains stuck from the start, regresses, or sputter starts with no momentum. The researcher is left, like Gregory Bateson, to invent a form that exemplifies the kind of conversation they hope to catalyze and analyze. It is no coincidence that Brad's innovation of improvisational therapy coincided with his development of RFA (Keeney, 1990, 1991).

While RFA can map how a conversation does not enact change or detail how it almost moved forward but stalled, its pinnacle achievement is found in how it illustrates a skillfully executed conversation in which clear context change took place. Again we underscore that a well-formed recursive frame analysis requires the recursive duo of a conversationalist's tacit ability to conduct a transformative conversation and a researcher whose RFA analytical skill can make the transformation in that conversation explicit. More importantly, these two sides—a change-oriented performer and a research analyst with a keen sense of constructivist dynamics—mutually benefit one another when embodied by the same person. In summary, the study of Ouroborean process requires both an exemplification of Ouroborean form and method of analysis.

The performance of a transformational conversation informs and reforms a researcher to better observe, identify, and analyze what RFA is designed to make explicit. Unless transformative talk can be personally enacted, the investigator likely lacks an experiential reference for discerning its critical performance dynamics in others. The most capable RFA researchers are equally versed at engaging in transformative conversation. Our dual performance-and-analysis pedagogical orientation begins with enacting the kind of data RFA is designed to examine, doing so before analyzing other conversations. This approach fosters greater nuanced familiarity with the patterns RFA can uniquely detect and explicate. If you wish to map with RFA, then learn to enact the principles of RFA.

The Primary Distinction in RFA's Name Is *Recursive*

As inventors, practitioners, and teachers of RFA, we have found this method is often elusive for newcomers to conceptually grasp and equally slippery to apply. This is partly due to the higher order circularity of the term *recursive*. Past the simpler notions of linearity and circularity, recursion denotes movement, the re-entry of a circle into itself, symbolized by the mythical Ouroboros, the dragon hero of cybernetics that swallows its own tail. Each time the dragon subsequently reemerges it may be markedly different or bring more of the same, depending upon what it recently swallowed. An *Ouroborean conversation* similarly chases its own thematic tale and with each go around the conversation either further entrenches itself in the same theme, or it moves forward in a distinctly altered way. For change to occur, when something new is introduced into the conversational circle it must be picked up and re-indicated, that is, fed back into the conversation in a sufficiently meaningful and enriched way. As this process continues, the conversational dragon grows and gains momentum, experienced as a dramatically shifted reality.

Bateson preferred that his conversation embody the subject matter being discussed—hence, the *meta* of his metalogue. RFA was designed to make explicit whether such a conversation succeeds, partially succeeds, never gets off the ground, or becomes circularly generative and change bound. Bateson's explorations of innovative conversation became a compass setting for Mary Catherine Bateson as well as our own performance and analytical work. We suggest that its advancement is better served by leaping from the notion of meta to recursion, or from metalogues to Ouroborean conversations, the kind of change-oriented communication that alters all involved, especially the context holding their interaction. Here the double nature of any perspective or angle of understanding flips as it turns—observers become the observed, practitioners become the client and so forth, as noise becomes the signal, distinctions become frames, and frames become parts dissolved in a vaster whole.

Our previous descriptions of RFA typically began with introducing the simple three-act structure of a play used by playwrights and screenwriters. However, this linear track does not reveal the whole circular dynamic that constitutes the movement from one frame to the next. It is the interaction of the conversation with itself that makes talk come to life. What is pragmatically overlooked by those unaccustomed to examining the circular patterns of communication is that transformative conversation does not advance in a straightforward way. Even when talk can be mapped to appear that it is a linear progression, the means of conversational movement are circular; a conversation interacts with its own productions, just like a masterfully composed theatrical play. Mary Catherine Bateson noted this dual action in the historic conference she, early in her career, chronicled in Austria. There Warren McCulloch pointed out that any linear movement was due to oscillations moving it along the way—a zig-zagging oscillation of cybernetic self-correction. As Norbert Wiener knew from the first launch of cybernetics, you hit the target by constantly changing how you deviate from it, narrowing the margin of difference as things progress.

In a theatrical play, the content of the first act, including its characters, setting, primary themes, conflicts, and so on, are not entirely left behind as the play progresses. These elements are creatively woven into subsequently changed action scenes as the drama unfolds from one act to the next. This progression is accomplished through the effective use of an ongoing recursive dynamic that keeps all the elements in play as they change in the construction of shifting action scenes. This inclusion applies to the beautiful the ugly as well, as Mary Catherine Bateson (1991) reminds us of the higher aesthetic frame: "Part of the task of composing a life is the artist's need to find a way to take what is simply ugly and, instead of trying to deny it, to use it in the broader design" (p. 211). Difference in all its forms, including the ugly, wrong, and mistaken, is not included for altruistic concerns. It is valued as essential for making the unfolding drama—of a play, therapy session, or life—feel rich, authentic, and real.

The recursive movement of conversation from one act or frame to another requires the skillful art of back-and-forth communicational moves, experienced as juggling the expected and unexpected. This dynamic provides familiar stability while hinting at something significantly different. The conversation advances by rocking backward and forward on a fulcrum until sufficient recursive momentum topples the performers into a dramatic change of scene. Change is improvised, precipitated by unexpected metaphors, surprising spoken lines, and uncommon patterns of interruption and reorganization.

Developing an ear to hear the circular turns, recursive feedback loops, and interactionally harvested feedforwards in a conversation is exemplified by maverick family therapists whose performance skills replace sideline observation and narration. It is more difficult to master creative interactivity than it is to memorize a model based on prefabricated ideology. This requires the skill of improvisation, often missing in the training of most change-oriented conversationalists, including therapists. As Mary

Catherine Bateson (in Tippett, 2015, n.p.) noted, "improvisation is a high order of skill" and requires "practice … by the hour."

Discerning the circularity, interactivity, and recursivity of a transformative conversation requires that non-linear dynamics are a familiar experience rather than only theoretically imagined. RFA as a research method only springs to life in the hands of the performer familiar with its experiential terrain, analyzing conversational material that clearly moves from one frame to another. Otherwise, RFA becomes a reductionist method of spelling out simple story lines prone to interpretive indulgences, missing the dramatology of a live performance that turns lines into circles and re-indicated distinctions into expanding frames. RFA is a research method designed to reveal how transformational conversation is established through a recursively structured performance, and to demonstrate that the latter is what evokes a profoundly noticeable experience of change. This kind of verbal production changes the conversationalists as well as those later analyzing the exchange.

No Ending, No Beginning

The duo of improvisation and composition is like a Batesonian father and daughter metalogue in search of a vaster clarity of recursive mind, greater creative invention, more cybernetic changeability, and wilder vision of possibility. What outcome does the real time performer paired with the post hoc investigator seek in this cybernetic tale with no beginning or end? Is it a pragmatic result like a happy ending at the end of the linear trail? Or is it a new beginning like a whole reality reset, enabling life to start differently the next time around? We suggest that it is more than these, but not exclusive of either. Like the Batesons, let's together seek the aesthetics that awaken us to emptier, fuller, deeper, and higher participation with all our relations inside the whole ecology of mind in body (Keeney, 1983a; Keeney & Keeney, 2012). Rather than linger and ponder, let's keep moving toward highest orders of wonder where angels fear to tread but still wander.

References

Bateson, G. (1972). *Step to an ecology of mind*. New York: Ballantine.
Bateson, G. (1979). *Mind and nature: A necessary unity*. New York: E. P. Dutton.
Bateson, M. C. (2003). "Daddy, can a scientist be wise?" *The American Journal of Semiotics*, 19(1-4), 3–15.
Bateson, M. C. (1991) *Composing a life*. New York: Atlantic Monthly Press.
Bateson, M. C. (1994). *Peripheral visions: Learning along the way*. New York: Harper Collins.
Brand, S., Bateson, G., & Mead, M. (1976). For God's sake, Margaret: Conversation with Gregory Bateson and Margaret Mead. *CoEvolutionary Quarterly*, 10/21(Summer), pp. 32–44.
Bateson, M. C. (2005). *Our own metaphor: A personal account of a conference on the effects of conscious purpose on human adaptation*. Hampton Press.
Keeney, B. P. (1976) *On paradigmatic change: Conversations with Gregory Bateson*. Unpublished manuscript.
Keeney, B. P. (1979). Glimpses of Gregory Bateson. *Pilgrimage*, 7, 17—44.
Keeney, B. P. (1983a). *Aesthetics of change*. New York: Guilford Press.
Keeney, B. P. (1983b). Size and shape: The story of a conference," *Journal of Strategic and Systemic Therapies*, 2, 72–79.
Keeney, B. P. (1990). Un Metodo per Organizzaire la Converszione Psicoterapia. *Terapia Familiare*, 32, 25–39.
Keeney, B. P. (1991). *Improvisational therapy: A practical guide for creative clinical strategies*. New York: The Guilford Press.

Keeney, B. P., & Ross, J. M. (1985). *Mind in therapy: Constructing systemic family therapies.* New York: Basic Books.

Keeney, H. & Keeney, B. (2012). *Circular therapeutics: Giving therapy a healing heart.* Phoenix, AZ: Zieg, Tucker, and Theisen.

Keeney, H. & Keeney, B. (2019). *The creative therapist in practice.* New York: Routledge.

Keeney, H., Keeney, B., & Chenail, R. J. (2012). Recursive frame analysis: A practitioner's tool for mapping therapeutic conversation. *The Qualitative Report, 17*(38), 1–15. Retrieved October 26, 2021 from http://www.nova.edu/ssss/QR/QR17/rfa_keeney.pdf

Keeney, H., Keeney, B., & Chenail, R. (2015). *Recursive frame analysis: A qualitative research method for mapping change-oriented discourse.* Fort Lauderdale: The Qualitative Report Books. Retrieved October 26, 2021 from https://nsuworks.nova.edu/tqr_books/1

Lipset, D. (1982). *Gregory Bateson: The legacy of a scientist.* Boston: Beacon Press.

Spencer-Brown, G. (1969). *Laws of form.* London: Allen & Unwin.

Tippett, K. (2015, October 1). Living as an improvisational art [Streamed interview and transcript]. At *On Being* [website]. Retrieved October 26, 2021 from https://onbeing.org/programs/mary-catherine-bateson-living-as-an-improvisational-art

Lorusso, Mick. (2009). *Emanating Presence.* Urban Cosmos Series, Meditative Interventions. Found object photograph. 25 x 30 cm.

Lorusso, Mick. (2010). *Their Songs/Their Presence*. Essence of Light/Life Series, Energy Patterns. Drawing. Watercolor pencil on paper. 77 x 55 cm.

Cybernetics and Human Knowing. Vol. 28 (2021), nos. 3-4, pp. 85–101

Cybernetics, Global Issues, and the Need for Systemic Wisdom
Mary Catherine Bateson and the Salzburg Global Seminar

William J. Reckmeyer,[1] PhD

Although Mary Catherine Bateson is best known for the award-winning books she wrote as a cultural anthropologist throughout her nearly 60-year career, she also spent a good part of her final three decades as a world-class cybernetician who focused on the importance of cybernetics and systems thinking for addressing the complexity of modern human affairs. This included significant work on different aspects of global issues, but her contributions on that topic are not well known—even within the cybernetics community—because they were scattered throughout occasional writings and presentations. This article examines the most extensive and developed expressions of Bateson's views about these matters, which occurred in a series of talks she gave at the Salzburg Global Seminar and the SJSU Salzburg Program in 2007-2010. Bateson was primarily concerned about the unintended counterproductive effects of human activities. Her talks highlighted three key aspects of the global problematique where cybernetics could be most helpful and reinforced her belief that cybernetics offers a powerful means for helping people become more informed and engaged global citizens who are able to systemically tackle the complex issues facing humanity in the Anthropocene.
Key Words: Mary Catherine Bateson, cybernetics, global issues, Salzburg Global Seminar

Introduction

During the years that I worked closely with Mary Catherine Bateson [2005-2010], I was often impressed by her ability to craft her own legacy as a highly-regarded scholar when many people tended to focus on her lineage as the daughter of famous parents—Margaret Mead and Gregory Bateson—who were among the most influential intellectuals of the 20th century. Catherine was primarily a cultural anthropologist, as were both of her parents, and highly regarded for her own scholarly works as well as for several collaborative books with her father and the curation of her mother's works through the Institute for Intercultural Studies. But all three were more than well-known cultural anthropologists, however, and made significant contributions to a variety of other fields. Of particular relevance to our interests is that they were noted cyberneticians. Not only were they all in the room at ground zero when the field of cybernetics was created during the Macy Conferences, which were held in 1946-1953 (Von Foerster, Mead, & Teuber, 1950-1956; Pais, 2016; Brand, Bateson & Mead, 1976), when Gregory and Margaret were major participants and

---

1. Professor Emeritus | Leadership & Systems; Department of Anthropology, San José State University, San José, CA 95192-0113 USA. Email: william.reckmeyer@sjsu.edu.

Mary Catherine was a school-age child soaking up the emerging insights, but they also continued to be engaged with the field of cybernetics at various times during their respective careers.

In this article I want to explore the emergent features of several critical issues raised by Mary Catherine in her scholarly work that lie at the nexus of cybernetics and global issues. I do so for four major reasons. First—this topic is the foremost concern facing humanity in the 21st century, especially as we become more aware of the fundamental role human cyberneticity has played in creating the Anthropocene, and we should be using our expertise as cyberneticians to effectively address the topic. Second—the cybernetics community at large has disproportionately focused on theoretical matters at the expense of substantive topics like this, which haven't received the kind of attention they deserve, and we should improve the situation. Third—MCB's attention to different aspects of this topic are scattered throughout her writings and presentations over more than three decades, which has diluted their impact and made it difficult for people to appreciate her insights about the topic. Fourth—the most extensive and developed expressions of MCB's views about the topic occurred in a series of talks she gave at the Salzburg Global Seminar and the SJSU Salzburg Program [2007-2010], but they are generally unknown within the cybernetics community and I think colleagues would benefit from learning more about them.

Salzburg Global Seminar

Mary Catherine's involvement with Salzburg actually began in the summer of 1947, when she accompanied her mother as a seven-year-old to the first session of the Salzburg Seminar in American Civilization (Bateson, 1984). Mead had been invited to co-chair the six-week session, which was initiated by three Harvard students after they heard US Secretary of State George C. Marshall's commencement speech in 1946 when he announced a proposal for the Marshall Plan. Their goal was to create a "Marshall Plan for the Mind" that would introduce Europeans to the ideals of American liberal democracy so the horrors of World War II would never occur again (Hallman, 2017a). The session was held at Schloß Leopoldskron, a magnificent 18th-century palace (Prossnitz, Ryback, & Stegen, 1994) that was owned by Max Reinhardt [co-founder of the Salzburg Music Festival] and has been home to the Seminar ever since (Gallup, 1987). The session, which was led by 11 distinguished Seminar Faculty members, supported by 8 Harvard graduate students, and attended by 112 Seminar Fellows from 18 different European countries, was an intensive cross-cultural dialogue where participants collaboratively explored a variety of contemporary issues in the social sciences and humanities (Mead, 1947; Wright, 1947; Matthiessen, 1948; Smith, 1948; Eliot & Eliot, 1987; Ryback, 1997; Russon & Ryback, 2003).

The Seminar was important for several reasons that are germane to this article. For Bateson, it was her first trip abroad and an experience that marked "the beginning of her love affair with languages" (Bateson, 2008a, p. 1) and triggered an intense interest in other cultures that lasted the rest of her lifetime (Bateson, 2007c). For Mead, it provided her with a timely opportunity to test ideas she had learned from her anthropological studies in the South Pacific before they were cut off by World War II [and likely energized by her central role in the early days of the cybernetic revolution, since the Salzburg Seminar was held between the 3rd and 4th meetings of the Macy Conferences that year] by intentionally "turning them loose on the community of the Seminar itself" (Matthiessen, 1948, p. 60). For cybernetics, it was arguably a critical factor in the genesis of Mead's subsequent admonition in the late 1960s that the field should apply its own insights to itself—the cybernetics of cybernetics, which has come to be known as second-order cybernetics—as a primarily human-centric rather than a techno-centric enterprise (Mead 1968). For the Seminar, Mead's glowing report to the Harvard Student Council [which was the principal financer and co-sponsor of the session] strongly recommended that the Seminar should be held again and offered a detailed list of suggestions to strengthen its operations (Mead, 1947).

Although the Seminar was never intended to become a long-term institution, Mead's report proved influential. It precipitated follow-up sessions starting the following year (Gleason, 1949) and eventually led to its establishment as a highly regarded non-governmental organization that is presently headquartered in Washington, DC and operationally based at Schloß Leopoldskron in Salzburg, Austria (Hallman, 2017f; Ryback, 2021). Ever since its origins, the Salzburg Global Seminar has been convening an evolving variety of sessions to encourage the free exchange of ideas, diverse viewpoints, informed discussions, and shared understandings in a neutral setting. Seminar sessions always focus on timely topics, ranging from political and social affairs to economic and cultural matters, and typically attract a heterogeneous group of knowledgeable and well-placed participants from a broad variety of countries around the world. By the middle of the 1950s it had established itself as a formal non-profit educational institution, purchased Schloß Leopoldskron, started organizing a half-dozen shorter sessions on contemporary topics each year, and initiated the study of American Studies in Europe (Eliot & Eliot, 1987; Ryback, 1997; Hallman, 2017b; Salzburg Global Seminar, 2021b).

The Seminar expanded the topical and regional breadth of its programs in the early-1960s, when it began introducing a broader spectrum of international topics along with the recruiting of Fellows from Central and Eastern Europe. This was a major shift for the Seminar and was due to increased funding from a number of private foundations, notably the W. K. Kellogg Foundation (Overton, 1995; Hallman, 2017c). As tensions tightened during the height of the Cold War, the expanding nature of its curriculum and growing diversity of its participants led the Seminar to adopt a "Common Problems" approach rather than continuing its historic emphasis on helping Western Europeans learn about America (Hallman, 2017c). While this outreach was greeted with suspicion by foreign ministries in Eastern European capitals, it soon led to a substantial increase in Fellows from that part of the continent. As a result of that initiative, as well as its fortuitous location in a neutral country in the heart of Europe, the Seminar grew into one of the few forums in the world where professionals from both sides of the Iron Curtain could safely meet to frankly discuss sensitive topics off the record (Eliot & Eliot, 1987; Ryback, 1997; Hallman, 2017c).

The Seminar expanded the topical and geographical breadth of its programs again in the early 1990s, immediately following the end of the Cold War, when forward thinking by its new leadership prompted the Salzburg Seminar to develop a more comprehensive focus on the world at large. This strategic change included a major initiative to strengthen the capacity of higher education institutions in the former Soviet bloc, long-term projects to support the reforming economies of Eastern Europe and the emerging economies of Asia, and outreach efforts to the Global South in Africa and South America (Ryback, 1997; Hallman, 2017d). It also prompted the Kellogg Foundation to substantially increase its financial support so the Seminar could implement these and other programs. My involvement with the Seminar began during the mid-1990s, when I was invited as one of several Kellogg National Leadership Fellows to support the Foundation's efforts to help globalize the Seminar. This was principally due to my expertise in cybernetics and leadership, particularly my experience using systemic approaches to help policy makers develop integrated strategies for collaboratively addressing complex global issues (Reckmeyer, 1991, 1993, 2016a, 2016b; Reckmeyer & Reckmeyer, 1993).

After serving as a Fellow at several Salzburg sessions during the late 1990s, my involvement increased when the Salzburg Seminar developed its pioneering International Study Program in Global Citizenship in 2004. Supported by additional funding from the College Board and Kellogg Foundation, the ISP's mission was to enhance the ability of US higher education institutions to prepare global citizens—people who know how to live and work in an increasingly interconnected world and build a more sustainable planet. The ISP was markedly different than previous Seminar programs, though, because it was a partnership between the Seminar and participating colleges and universities that sent cohorts of their own students and faculty to attend sessions targeting their own institutional interests rather than catering to individuals who were selected by the Seminar and attended on their own (Fried, Goldman, & Schroeder, 2012; Hallman, 2017e). I was invited by Olin Robison [Salzburg Seminar President] and Jochen Fried [ISP Director] to become part of the ISP team in 2004, initially as a Core Faculty Member and then as ISP Faculty Chair two years later. I also worked closely with Dr. Fried and his colleague David Goldman on recruiting other colleges and universities to join the ISP partnership, which eventually included the development of global citizenship programs on more than a half dozen college campuses in California that collectively sent nearly 400 faculty and student to week-long sessions each summer during the twelve years of the ISP Program.

One of the most robust results of those efforts was the SJSU Salzburg Program at San José State University, which is the leading public institution of higher education in Silicon Valley. The purpose of that Program, which I co-founded and directed from 2005-2017, was to develop a critical mass of change agents from across our campus to help globalize our university and educate globally competent citizens. Each year we selected 10-30 SJSU Salzburg Fellows [faculty and administrators] and SJSU Salzburg Scholars [students] to participate in an intensive 18-month program, funded them to attend an ISP session, and then worked with them intensively in the following academic year on a mix of self-organized projects (Reckmeyer, 2010). We also hosted a dozen ISP Faculty members and Salzburg Fellows as SJSU Distinguished Visiting Scholars; sent several SJSU professors to serve as ISP faculty members; hosted Dr. Fried as a Fulbright Visiting Scholar for a semester; co-organized and hosted a major Salzburg Global Fellows Seminar on "The Rule of Law in the International Community;" established an annual Peter Lee Memorial Lecture on global citizenship; co-sponsored three semester-long Provost's Honors Seminars on global citizenship and US national security policy in a global world; and undertook a variety of other supporting activities (Reckmeyer, 2010, 2011; Brown, 2011; Reade, Reckmeyer, Cabot, Jaehne, & Novak, 2013). As a result of these efforts, the Program was honored as a *Top 10 Program in Global Citizen Diplomacy* by NAFSA, the United States Center for Citizen Diplomacy, and the United States Department of State (Harris, 2010).

Global Issues

Mary Catherine's subsequent involvement with Salzburg in 2007 began when I reconnected with her at the 2005 Annual Conference of the American Society for Cybernetics in Washington, DC. We were both giving presentations that weekend. Hers was a major keynote address on the importance of active wisdom (Bateson, 2005d), which became the core of her next book, and was when she first mentioned her idea of "Adulthood II" (Bateson, 2010b, p. 13). Mine was a presentation on the work that I had been leading on systems-of-systems approaches to public policy making and program management for the US government (Reckmeyer, 2005). I hadn't seen her in a while, so we took a few moments to catch up. I told her what I was doing with the ISP and asked whether she might like to join us in Salzburg. She was very interested. It was partly because she had so many good memories of her summer there in 1947 and had always wanted to return, but also partly because she was intrigued by what we were doing at the ISP in terms of the intersection between learning and global citizenship. The timing was perfect because the Seminar was just starting to make plans for celebrating its 60th anniversary in 2007, so I arranged for her to spend some time at Schloß Leopoldskron that summer.

As a result of that invitation, Bateson delivered a set of six Salzburg-related talks between 2007 and 2010. These included her keynote address for the Seminar's 60th Anniversary (Bateson, 2007b); formal presentations at four ISP sessions (Bateson, 2007a, 2007d, 2008b, 2010a); and her Peter Lee Memorial Lecture for the SJSU Salzburg Program (Bateson, 2010c). Most of those talks were not isolated performances, other than her keynote address, but were part of week-long educational activities where she actively engaged in formal exchanges and informal conversations with participants. Although I am not at liberty to share details about her talks at the four ISP sessions, since they were conducted under Chatham House rules to ensure complete freedom of expression [a long-time hallmark of the Salzburg experience that had begun with Session #1 in 1947], the points she made in those sessions were expressed more explicitly and fully in her two public talks. All of her presentations included stories about the historical context for her comments, starting with her memories of Schloß Leopoldskron after WWII, and continued with observations about key global developments that had occurred during the ensuing decades. They also included more extensive comments about two critical aspects of global issues—the "diversity of cultures and … the possibility of environmental disaster" (Bateson, 2008a, p. 1)—that were consistent themes in her presentations at Salzburg and San José State, as well as other instances during the following decade (Bateson, 2011, 2015c, 2016b, 2018).

Bateson was adept at using stories and metaphors to illuminate deeper insights, an ability to intuitively perceive important patterns and similarities that Aristotle in his Poetics considered a "sign of genius" (Barnes, 1984, p. 2335), so it seems appropriate to share a metaphor that she and I agreed captures the essence of these two themes. The metaphor emerged during a long chat that we had on the day after her keynote. We were sitting on the Schloß Leopoldskron terrace, overlooking the lake and the Untersberg massif that lies at the northernmost edge of the Austrian Alps—a vista that is familiar to millions of people as the setting where Maria and the von Trapp children fell in the water during the *Sound of Music*, which is still the most popular film in history [based on paid attendance] and remains the primary driver of tourism in Salzburg more than fifty years after it was released. MCB observed that many important lessons for life often appear at an early age, such as learning to play together in the sandbox, a metaphor I knew she had told before in other settings. I responded—after all, it was a conversation—by adding that it was also where people can learn to take care of the sandbox, so we have a place to play another day (Bateson, 2007c). Both of these lessons reflect fundamental insights, but unfortunately are ones that many people and humanity as a whole don't seem to have learned very well yet.

Neither of these themes about global issues writ large—how people interact with each other and with the natural world we inhabit—were central to most of Mary Catherine's published work, but she did note that her experience in Salzburg had initiated a "new sense of commitment" (Bateson, 2008a, p. 1) to address broader issues like the "ecology of the planet, or the climate, or large populations of human beings that have evolved for many years in separate locations and are now re-integrating" (Bateson, 2018, n.p.). There were a few examples, including *Our Own Metaphor* and several articles, that examined selected aspects of these themes (Bateson, 1990, 2005e, 2007; Goldsby & Bateson, 2019). However, none of them emphasized the broader human condition as explicitly and deeply as she did in her Salzburg-related talks. Moreover, none of them focused on cybernetics or its relevance to these core themes. Bateson's non-technical presentations were particularly effective because her audiences at the talks were primarily composed of faculty members, administrators, staff, or students who were not experts in global issues and knew even less about cybernetics and the value of systems thinking. But they were interested in exploring global citizenship and how it might enhance their learning, teaching, research, and administration.

The topic of global citizenship was not even on the radar screens of most academics or the public at large when the Seminar launched the ISP in 2004. It was not an entirely new idea, since its roots can be traced back to ancient Greece (Schattle, 2009), but it started attracting attention as people became more conscious about the impact of globalization. Although a good case can be made that the world has been globalizing for much longer than most people realize (Baldwin, 2016; Marks, 2007; Osterhammel & Petersson, 2003; Schwab, 2016; Sachs, 2020; Smil, 2021), globalization didn't start becoming a popular term until the 1990s. This was chiefly due to two major developments that occurred in 1991—the end of the Cold War and the creation of the Internet. The synergistic effects of those events launched an expanding set of global connections that pervade virtually every aspect of modern life. The result is that humanity is now in the midst of evolving from an international world [where interactions had been constrained and shaped by national governments since the Treaty of Westphalia in 1648] into a global world [where inter-actions are increasingly self-organizing on both an individual and a collective basis]. For the first time in human history, it is possible to describe the world as a highly integrated global community of human beings in actual rather than metaphorical or aspirational terms. It also highlights the lack of appropriate institutions to regulate these evolving interactions.

There are many upsides to these emerging conditions, but there are also many downsides. The expanded scope, intensity, velocity, and impact of human interconnections in every nook and cranny of the world has been producing an extraordinary combination of global issues that are best described as the Anthropocene (McNeill & Engelke, 2014; Steffen, Broadgate, Deutsch, & Gaffney, 2015; Kress & Stine, 2017; Lewis & Maslin, 2018). In my view, these issues include two basic sets of substantive challenges that have been generated by a dramatic increase in humanity's cyberneticity over the past 500 years—sophisticated capabilities to transform ourselves and our world through effective purposeful agency that have become so significant I think it is more accurate to view the species as *Homo Cyberneticus* rather than *Homo Sapiens* (Reckmeyer, 2019, 2020, 2021). Some of those challenges are people-centric [nuclear war, cyber terrorism, economic inequality, racial injustice, and cultural tribalism], while the others are planet-centric [climate change, pandemics, species extinction, depletion of resources, and the nitrogen cycle]. Each one of them is a major threat to our collective well-being and all of them are wicked issues (Rittel & Webber, 1973; Conklin, 2008), although I think they are better understood as complex messes that need to be managed rather than well-defined problems that can be solved (Ackoff, 1974). Some of these challenges may even pose existential risks, but every one interacts with the others in ways that are jointly jeopardizing the habitability of Planet Earth—at least for human civilization as we know it (Meadows, Randers, Meadows, & Behrens, 1972; Bostrom & Cirkovic, 2008; Smil, 2012; Rockstrom & Klum, 2015; Steffen et al., 2015; Raworth, 2017).

None of these factors were as appreciated when the ISP was launched as they are now, but there was a growing belief that the world was changing profoundly. These evolving conditions are not only creating new global realities, but they are also generating new global rules for how to succeed in these realities. The Seminar developed the ISP because Robison and Fried realized that most people didn't know how to live and work in such an interdependent world, given its unprecedented nature, so the ISP was designed to promote and facilitate new kinds of thinking and acting. Each ISP session, and the Program as a whole, focused on the importance of using interdisciplinary sets of relevant experts to provide the same type of cross-national education and cross-cultural communication that has been a hallmark of the Salzburg Global Seminar since its inception. The critical difference now is that the world is truly global, rather than regions of loosely connected cultures. Now more, than ever, we need to educate global citizens who are both informed about and engaged with addressing today's critical issues (Reckmeyer, 2010, 2013) to help humanity to build a better and more sustainable world (Dower & Williams, 2002; Schattle, 2008; Schwab, 2008; Stearns, 2009; Altinay, 2011; Cabrera & Unruh, 2012; Green, 2012). This has also led to the emergence of numerous global citizenship education programs around the world (UNESCO, 2021; Oxfam, 2021; Global Citizenship Alliance, 2021; Global Citizen, 2021; Global Citizens Initiative, 2021; Global Citizen Year, 2021).

Cybernetics

All six of Bateson's talks at the Salzburg Global Seminar and SJSU Salzburg Program, like those she gave at the various ASC conferences where I heard her speak (Bateson, 2005d, 2014a, 2014b), emphasized stories and metaphors rather than the technical terminology that is typical of many cyberneticians and systems scientists. This not only reflected her expertise and experience as a cultural anthropologist, but also her belief that they offered powerful ways to illuminate complex points for a broad mix of audiences (Bateson, 2018). She made a conscious effort to place her Salzburg comments in the context of the Schloß and the arc of human history since World War II. A particularly vivid one was sharing her memory of attending a performance of the *Jedermann* play in 1947, which was (and still is) the highlight of the annual world-famous Salzburg Music Festival (Gallup, 1987), because it was a drama about the human condition that underscored the risks affecting "the health of the entire planet and the peace of the entire human species" (Bateson, 2007b). The overarching meta-theme of all her Salzburg-related talks was her concern about the growing disconnect between humanity and nature – a topic she had explored in her account of the conference organized by her father in 1968 [which was held at another Austrian castle, Burg Wartenstein] that examined the unintended consequences of human purposefulness (Bateson, 2005e), but which she had only started addressing in her own work shortly before she became involved with the ISP Program in 2007.

Mary Catherine's concerns about the state of the world had led her to revisit the importance of cybernetics and systems theory, which she hadn't done "for quite a long time" (Bateson, 2018, n.p.). She thought the field was essential for helping people address the challenges of complex wholes more effectively and responsibly. This led to an expansion of her scholarly interests, which was also reinforced by invitations to become more involved with the American Society for Cybernetics (Bateson, 2007c, 2018). Bateson didn't refer to cybernetics in her Salzburg-related talks, however, but always referred to systems theory—primarily because she had strong reservations about the evolution of cybernetics since the Macy Conferences. The tragedy of the cybernetic revolution, in her view, was that it had grown to emphasize narrow technological features rather than broader systemic features that characterized human and ecological conditions (Bateson, 2007c, Bateson, 2018)—a development that has been examined by several chronicles of the field (Heims, 1991; Hayles, 1999; Conway & Siegelman, 2005; Collins, 2007; Pickering, 2011; Kline, 2015; Chaney, 2017; Clarke, 2020). In each of her talks, but most pointedly in her Peter Lee Memorial Lecture, Mary Catherine was chiefly interested in the unintended counterproductive effects of human activities and focused on three principal aspects of the global problematique where she thought cybernetics could be most helpful.

The first aspect was the need for an integrated appreciation of the world that emphasizes a more harmonious balanced relationship between people and our planet. This was a concern that echoed Bateson's closing thoughts about the Burg Wartenstein conference, when she recognized the value of creating "a unified and widely shared vision, a vision that would be persuasive both intellectually and emotionally, to provide the context for action" (Bateson, 2005c, p. 314). She believed that there had been a growing sense of common dangers in recent decades and felt that we were experiencing a new time of crisis in human history, but she also believed that there was no common sense of why these dangers were escalating—much less any common sense of what should and/or could be done to resolve them in an effective and ethical way (Bateson, 2010c). The source of the dilemma for Mary Catherine, as it was for her father, was the extreme emphasis on individual independence that has become characteristic of the WEIRD [Western, Educated, Industrialized, Rich, Democratic] cultures that have created the modern world (Henrich, 2020). Her recommendation was to recognize that we live in a highly interdependent world where the real "unit of survival is a flexible organism-in-its-environment" (G. Bateson, 1970/2000b cited in Bateson, 2016b, p. 2) and to emphasize our common benefits rather than our individual interests.

The second aspect was the need to collaboratively address critical substantive issues in a timely way, especially if we are able to develop the kind of unified and shared vision that she was advocating. There are two factors that make this particularly difficult in a world of nearly 8 billion people. One factor includes the tensions generated by a systemic tragedy of the commons (Hardin, 1968), which highlight the increasing difficulty of integrating individual and collective priorities (Bateson, 2016c; Haidt, 2012) when humanity has developed sophisticated cybernetic capabilities that enable members of the species to increasingly maximize their own self-interests. The other factor includes the tensions generated by the positive and negative effects of even the most well-intentioned of actions. Bateson's parents exemplified this tension, as she noted about their responses to their experiences during World War II and the Macy Conferences. Her father was pessimistic—interested in the intellectual possibilities of cybernetics, but suspicious about engaging in conscious purposeful behavior given how counterproductive it has been. Her mother was optimistic—interested in the practical possibilities of cybernetics for improving human relations and believing in humanity's capacity to intentionally build a common future through the judicious application of its capabilities (Bateson, 2005a, 2010c; Brockman, 2021).

The third aspect was the need to strengthen humanity's capacity for learning how to achieve both of these recommendations, specifically in terms of using cybernetics to help humanity think and act more systemically. Bateson knew that educating people to become better global citizens wouldn't be easy, given the serious challenges of the issues themselves and the equally serious challenges of working together to address those issues. Drawing on research from her research over the prior two decades (Bateson, 1994, 2004), she emphasized that human beings were capable of profound learning, to a much greater extent than has been evidenced by any other form of life, and that it is this capability that offers the greatest hope for building a better world (Bateson, 2010c). Although we never explicitly discussed the topic, I think she might agree with my view that this is the most transformational feature of humanity's cybernetic capabilities—a meta-capability that can amplify everything people do. This is not only because it enables us to engage in proto-learning [learning specific knowledge and skills] and deutero-learning [learning how to learn], critical distinctions that her father originally noted even before the Macy Conferences (G. Bateson, 1942/2000a), but because it also enables us to engage in the reflexive-learning [learning about ourselves and the contexts of our actions] that is essential for addressing the global challenges facing humanity and our world in the 21st century. The role of human learning was so important to Mary Catherine that she had adopted it as a personal slogan—"We are not what we know, but what we are willing to learn"—shortly before her first set of talks in Salzburg (Bateson, 2007c).

In Bateson's view, cyberneticians and educators have a responsibility to create contexts in which people experience interactions and relationships that prepare them to see the world differently (Bateson, 2010c). Otherwise, we will continue generating insanity traps of our own making where people engage in the same purposeful behaviors with the expectation of achieving better results. Mary Catherine was convinced that building a more sustainable and equitable world requires more than good will and technical knowledge. It requires systemic wisdom and the ability to engage in interdependent cooperation based on a sophisticated appreciation of the bigger picture. In a complex modern global world increasingly characterized by black swan events that are inherently unpredictable challenges with potentially catastrophic consequences (Taleb, 2010) and gray rhino events that are highly probable and known challenges with potentially large impacts that we tend to ignore (Wucker, 2016), humanity cannot rely on traditional approaches that tend to treat people as separate beings and our world as a resource to plunder as we wish (Bateson, 2016b). In her Salzburg talks, Bateson argued that the signs are becoming abundantly clear—humanity has been unintentionally producing a planet-wide crisis, despite the undeniable improvements in human conditions that began escalating during the past 500 years, and we're running out of time to improve the situation.

Bateson offered two closing pieces of advice in those talks about what needed to be done. One point started with a warning—that the greatest fear from climate change and other global issues was that people tend to get very nasty when they think their share of the pie will be getting smaller and that we're going to face incredible violence similar to what happened in Yugoslavia if we don't start behaving differently (Bateson, 2010c). Her advice was that it would be vital to "create a global system of sharing and cooperation … and a world that people feel is just, one where they understand that if we cooperate there is more, that we do not live in a zero-sum game" (Bateson, 2010c, n.p.). The other point offered a ray of hope—that the dramatic increases in life expectancy among developed populations of the world over the past century, which had led to the emergence of entire cohorts of retired people with extensive knowledge and experience, represented a huge opportunity for our species (Bateson, 2010b). Her advice was that people should learn more about the nature of active wisdom, which she envisioned as flowing from the combination of experience and continuing health, and encourage those living in what she called the Adulthood II phases of their lives to actively share and apply their wisdom before it was too late (Bateson, 2010c, 2015a). Both pieces of advice reflected her conviction that we all have the capacity to contribute, but we don't always know how to do so. They also reinforced her points throughout these talks that cybernetics offers a powerful means for helping people become more informed and engaged global citizens who can systemically tackle the complex issues facing humanity.

Conclusion

Although Mary Catherine's Salzburg-related talks began as a "sentimental journey" to revisit Schloß Leopoldskron (Bateson, 2008a, p. 1), they amplified her growing interest in helping people learn about cybernetics and its relevance to global issues. Her talks were also influenced by the book she was writing during those three years about active wisdom (Bateson, 2010b), which led her to increasingly share her own evolving wisdom about the challenges of the Anthropocene during the final years of her life. Bateson was a prolific author, most of whose work was anthropological in nature, but her contributions to cybernetics and global issues were a significant part of her overall body of work and shouldn't go unappreciated even though much of it was shared in talks and occasional articles rather than in books. Notable examples include the letter she wrote to the American Society for Cybernetics in response to receiving the Norbert Weiner Gold Medal (Bateson, 2011); keynote addresses at the ASC's 50th

Anniversary Conference (Bateson, 2014a, 2014b), the ISSS Conference on Governing the Anthropocene (Bateson, 2015a), and the IEEE Conference on Norbert Wiener in the 21st Century (Bateson, 2015c); as well as a couple of articles (Bateson, 2016b, 2016c). Each of them focused on her concerns that we are living in a time "of great danger, but we are unable to discuss it in ways the public can grasp" (Bateson, 2015c, p. 35)—and that we'll never be able to develop the coordinated responses significant global challenges warrant if we experts don't figure out how to communicate better.

I don't think it is possible to fully appreciate Mary Catherine's legacy without understanding it in the context of her own life, a life that was certainly framed and influenced by her role as the only daughter of famous parents who used cybernetics to address serious global issues in their own ways (Bateson, 2005e; Kellogg & Mead, 1976). But it would be a serious mistake to assume that Mary Catherine's contributions were only derivative, for she was a world-class cybernetician who carved her own noteworthy path despite the long shadows cast by both of her parents. Although better known as a distinguished cultural anthropologist (Green, 2021; Schudel, 2021), Mary Catherine exemplified the kind of responsible competence that Heinz von Foerster had advocated in the early days of the cybernetics field when he urged his colleagues to "fulfill their social and individual responsibilities as cyberneticians who should practice what they preach" (Von Foerster, 1972, p. 197). The nexus of cybernetics and global issues was very important to Mary Catherine (Bateson, 2007c; Brockman, 2021; Salzburg Global Seminar, 2021a), as it has been to me throughout my career (Reckmeyer, 2016b), even though neither topic became a focus for Bateson until the latter portion of her career. She also greatly appreciated her interactions with the ASC and Salzburg communities, as she shared in conversations with me and her daughter (Bateson, 2010d; Kassarjian, 2021) in the years since her Salzburg-related talks, in addition to noting how rare it was to participate in vibrant groups who cared so deeply about such matters (Bateson, 2011). Considering the wake-up calls humanity has been experiencing in recent years, political and pandemical, I'm hoping that her work and her spirit will inspire more cyberneticians and systems thinkers to focus their attention and talents on addressing critical global issues before *Homo Cyberneticus* destroys the systems that are essential to our well-being.[2]

---

2. On a related note, given the Salzburg Global Seminar's long history of transnational and transdisciplinary interests since its founding in 1947 (the Seminar will celebrate its 75th anniversary in 2022), I've always found it surprising that so few cyberneticians and systems scientists have participated in any of its sessions. Besides Mead (1947), Bateson (2007-2008, 2010), and myself (1995, 1997, 1998, 1999, 2004-2015), others included Kenneth Boulding (1949, 1959), Norbert Wiener (1962), and Stuart Umpleby (2002).

Mary Catherine Bateson at the Salzburg Global Seminar
(Photo: Salzburg Global Seminar, 2007-07-14)

Schloß Leopoldskron—Home of the Salzburg Global Seminar
(Photo: William J. Reckmeyer, 2008-07-16)

References

Ackoff, R. L. (1974). *Redesigning the future: A systems approach to societal problems.* New York: John Wiley.
Altinay, H. (2011). *Global civics: Responsibilities and rights in an interdependent world.* Washington, DC: Brookings Institution Press.
Baldwin, R. (2016). *The great convergence: Information technology and the new globalization.* Cambridge, MA: Belknap Press of Harvard University Press.
Barnes, J. (Ed.). (1984). *The complete works of Aristotle—Volume two: Poetics.* Princeton, NJ: Princeton University Press.
Bateson, G. (2000a). Social planning and the concept of deutero-learning. In *Steps to an ecology of mind: Collected essays in anthropology, psychiatry, evolution, and epistemology* (pp. 159–176). Chicago: University of Chicago Press. (Originally published in 1942)
Bateson, G. (2000b). Form, substance, and difference. In *Steps to an ecology of mind: Collected essays in anthropology, psychiatry, evolution, and epistemology* (pp. 454–471). Chicago: University of Chicago Press. (Originally published in 1970)
Bateson, M. C. (1984). *With a daughter's eye: A memoir of Margaret Mead and Gregory Bateson.* New York: William Morrow & Company.
Bateson, M. C. (1990). Beyond sovereignty: An emerging global civilization. In R. B. J. Walker & S. H. Mendlovitz (Eds.), *Contending sovereignties: Redefining political communities* (pp. 196–220). Boulder, CO: Lynn Reinner Publisher.
Bateson, M. C. (1994). *Peripheral visions: Learning along the way.* New York: Harper.
Bateson, M. C. (1996). Democracy, ecology, and participation. In R. Soder (Ed.), *Democracy, education, and the schools* (pp. 69–86). San Francisco: Jossey-Bass.
Bateson, M. C. (2004). *Willing to learn: Passages of personal discovery.* Lebanon, NH: Steerforth Press.
Bateson, M. C. (2005a). The double bind: Pathology and creativity." *Cybernetics and Human Knowing, 12*(1), 11–21.
Bateson, M. C. (2005b). Foreword 1991. In *Our own metaphor: A personal account of a conference on the effects of conscious purpose on human adaptation* (pp. ix–xv). Cresskill, NJ: Hampton Press.
Bateson, M. C. (2005c). Afterward 2005. In *Our own metaphor: A personal account of a conference on the effects of conscious purpose on human adaptation* (pp. 313–324). Cresskill, NJ: Hampton Press.
Bateson, M. C. (2005d, October 27). Relationships between demographic changes and cultural transmission. [Keynote address]. *2005 Annual Conference of the* American Society for Cybernetics, October 27, 2005 in Washington, DC.
Bateson, M. C. (2005e). *Our own metaphor: A personal account of a conference on the effects of conscious purpose on human adaptation.* New York: Alfred A. Knopf. (Originally published in 1972)
Bateson, M. C. (2007a). *Educating for social and global responsibility.* Presentation at ISP #20: Faculty Session on Colleges and Universities as Sites of Global Citizenship at the Salzburg Global Seminar, July 13, 2007.
Bateson, M. C. (2007b). *Reflections and observations.* [Keynote address]. 60th Anniversary Celebration of the Salzburg Global Seminar, July 15, 2007.
Bateson, M. C. (2007c). Private conversation. At the Salzburg Global Seminar, July 16, 2007.
Bateson, M. C. (2007d). *Educating for social and global responsibility.* Presentation at ISP #20: Faculty session on Colleges and Universities as Sites of Global Citizenship at the Salzburg Global Seminar, July 20, 2007.
Bateson, M. C. (2007e). Education for Global Responsibility. In S. C. Moser & L. Dilling (Eds.), *Creating a climate for change: Communicating climate change and facilitating social change* (pp. 281–291). Cambridge, UK: Cambridge University Press.
Bateson, M. C. (2008a). Salzburg Seminar Celebrates 60 Years. *Notes From the Field —Newsletter of the IIS.* New York: Institute for Intercultural Studies.
Bateson, M. C. (2008b). *Educating for Social and Global Responsibility.* Paper Presented at ISP #26: Student Session on Global Citizenship and the World at the Salzburg Global Seminar, June 2, 2008.
Bateson, M. C. (2010a). *Living longer: Planetary consciousness and global citizenship.* Paper Presented at ISP #46: Faculty Session on *Colleges and Universities as Sites of Global Citizenship* at the Salzburg Global Seminar, July 14, 2010).
Bateson, M. C. (2010b). *Composing a further life: The age of active wisdom.* New York: Vintage Books.
Bateson, M. C. (2010c). *Global citizenship and the age of active wisdom.* Peter Lee Memorial Lecture for the SJSU Salzburg Program at San José State University, San José, CA, November 10, 2010.
Bateson, M. C. (2010d). Private conversation. At the SJSU Salzburg Program at San José State University, San José, November 7, 2010.
Bateson, M. C. (2011). Letter on Receiving the Norbert Wiener Gold Medal. American Society for Cybernetics, dated August 27, 2011. Retrieved on June 17, 2018 from https://www.asc-cybernetics.org/2011/?p=1571&_ga=2.33664041.980861917. 1630856451-557069978.1626191493
Bateson, M. C. (2014a). *Living in cybernetics.* [Keynote address]. American Society for Cybernetics 50th Anniversary Annual Conference. Washington, DC, August 6, 2014 Retrieved October 26, 2021 from https://www.youtube.com/watch?v=wpjnVVWXZMs&t=14s

Bateson, M. C. (2014b). *The future of cybernetics.* [Plenary address]. American Society for Cybernetics 50th Anniversary Annual Conference. Washington, DC, August 9, 2014. Retrieved October 26, 2021 from https://www.youtube.com/watch?v=nXQraugWbjQ&t=2s

Bateson, M. C. (2015a). *New world of active wisdom.* TEDx Talk, Cape May, NJ, January 12, 2015. Retrieved October 26, 2021 from https://www.youtube.com/watch?v=YIfrAjkc83w&t=453s.

Bateson, M. C. (2015b). *Causality and responsibility.* [Keynote address]. International Society for the Systems Sciences Annual Meeting on Governing the Anthropocene. Berlin, August 4, 2015.

Bateson, M. C. (2015c). Norbert Wiener: Odd man ahead. *IEEE Technology and Society Magazine, 34*(3), 35–36.

Bateson, M. C. (2016a). Living in cybernetics: Making it personal. *Cybernetics and Human Knowing* 23(1), 96–102.

Bateson, M. C. (2016b). The myths of independence and competition. *Systems Research and Behavioral Science, 33,* 674–677.

Bateson, M. C. (2018). How to be a systems thinker: A conversation with Mary Catherine Bateson. *Edge* [Website]. Retrieved October 26, 2021 from https://www.edge.org/conversation/mary_catherine_bateson-how-to-be-a-systems-thinker

Bostrom, N., & Cirkovic, M. M. (2008). *Global catastrophic risks.* Oxford, UK: Oxford University Press.

Brand, S., Bateson, G., & Mead, M. (1976). For God's sake, Margaret: Conversation with Gregory Bateson and Margaret Mead. *CoEvolutionary Quarterly,* 10/21(Summer), pp. 32–44.

Brockman, J. (2021, January 18). Mary Catherine Bateson: Systems thinker. *Edge* [Website]. Retrieved October 26, 2021 from https://www.edge.org/conversation/mary-catherine-bateson.

Brown, I. (2011). *Global citizenship and the international rule of law—Report.* Salzburg: Salzburg Global Seminar.

Chaney, A. (2017). *Runaway: Gregory Bateson, the double bind, and the rise of ecological consciousness.* Chapel Hill, NC: University of North Carolina Press.

Clarke, B. (2020). *Gaian systems: Lynn Margulis, neocybernetics, and the end of the Anthropocene.* Minneapolis, MN: University of Minnesota Press.

Collins, S. G. (2007). Do cyborgs dream of electronic rats? The Macy conferences and the emergence of hybrid multi-agent systems. In G. P. Trajkovski & S. G. Collins (Eds.), *Emergent agents and socialities: Social and organizational aspects of intelligence: Papers from the AAAI Fall Symposium on Emergent Agents and Socialities* (pp. 25–34). Arlington, VA: Association for the Advancement of Artificial Intelligence.

Conklin, J. (2006). Wicked problems and social complexity. In J. Conklin (Ed.), *Dialogue mapping: Building shared understanding of wicked problems* (pp. 1–19). New York: John Wiley.

Conway, F., & Siegelman, J. (2005). *Dark hero of the information age: In search of Norbert Wiener, the father of cybernetics.* New York: Perseus.

Dower, N., & Williams, J. (Eds.). (2003). *Global citizenship: A critical introduction.* New York: Routledge.

Eliot, T. H., & Eliot, L. J. (1987). *The Salzburg Seminar: The first forty years.* Ipswich, MA: Ipswich Press.

Fried, J. Goldman, D., & Schroeder, A. (2012). *International study program on global citizenship.* Salzburg, Austria: Salzburg Global Seminar.

Gallup, S. (1987). *A history of the Salzburg Festival.* London: Weidenfeld and Nicolson.

Gleason, H. (1949, September 26). Student council-sponsored Salzburg Seminar explains American civilization to Europeans. *The Harvard Crimson.* Retrieved October 26, 2021 from https://www.thecrimson.com/article/1949/9/26/student-council-sponsored-salzburg-seminar-explains/.

Global Citizen (2021). Accessed on September 1, 2021 at https://www.globalcitizen.org/en/.

Global Citizen Year (2021). Accessed September 1, 2021 at https://www.globalcitizenyear.org/content/global-citizenship/.

Global Citizenship Alliance (2021). Accessed August 1, 2021 at https://globalcitizenshipalliance.org/.

Global Citizens Initiative (2021). Accessed September 1, 2021 at https://www.theglobalcitizensinitiative.org/.

Goldsby, R. A., & Bateson, M. C. (2019). *Thinking race: Social myths and biological realities.* Lanham, MD: Rowman & Littlefield.

Green, M. (2012). *Global citizenship: What are we talking about and why does it matter?* Paper 4 of series: *Trends and Insights for International Education Leaders.* Washington, DC: NAFSA (Association of International Educators). Retrieved October 26, 2021 from https://www.researchgate.net/publication/265438069_Global_Citizenship_What_Are_We_Talking_About_and_Why_Does_It_Matter

Green, P. (2021, January 14). Mary Catherine Bateson dies at 81—Anthropologist on lives of women. *New York Times.* Retrieved January 16, 2021 from https://www.nytimes.com/2021/01/14/books/mary-catherine-bateson-dead.html.

Haidt, J. (2012). *The righteous mind: Why good people are divided by politics and religion.* New York: Pantheon Books.

Hallman, L. (2017a). *A Marshall Plan for the mind—1947.* Salzburg, Austria: Salzburg Global Seminar. Retrieved July 15, 2021 from https://www.salzburgglobal.org/news/latest-news/article/a-marshall-plan-for-the-mind-1947.

Hallman, L. (2017b). *From idealist experiment to eminent institution—1948 to 1961.* Salzburg, Austria: Salzburg Global Seminar. Retrieved July 15, 2021 from https://www.salzburgglobal.org/news/ latest-news/article/from-idealist-experiment-to-eminent-institution-1948-to-1961.

Hallman, L. (2017c). *Cold war crossroads—1962 to 1989.* Salzburg, Austria: Salzburg Global Seminar. Retrieved July 15, 2021 from https://www.salzburgglobal.org/news/latest-news/article/cold-war-crossroads-1962-to-1989.

Hallman, L. (2017d). *A globalizing world—1990 to 2004.* Salzburg, Austria: Salzburg Global Seminar. Retrieved July 15, 2021 from https://www.salzburgglobal.org/news/latest-news/article/a-globalizing-world-1990-to-2004.

Hallman, L. (2017e). *People and power—2005 onwards*. Salzburg, Austria: Salzburg Global Seminar. Retrieved July 15, 2021 from https://www.salzburgglobal.org/news/latest-news/article/people-and-power-2005-onwards

Hallman, L. (2017f). *Seven insights for seven decades*. Salzburg, Austria: Salzburg Global Seminar. Retrieved July 15, 2021 from https://www.salzburgglobal.org/news/latest-news/article/seven-insights-for-seven-decades.

Hardin, G. (1968). The tragedy of the commons. *Science, 162* (13 December), 1243–1248.

Harris, P. L. (2010). San José State selected for national honor recognizing Top Citizen Diplomacy Programs: SJSU Salzburg program to serve as model for higher education globalization. San José, CA: San José State University, SJSU Newsroom. https://blogs.sjsu.edu/newsroom/2010/san-jose-state-selected-for-national-honor-recognizing-top-citizen-diplomacy-programs/.

Hayles, N. K. (1999). *How we became posthuman: Virtual bodies in cybernetics, literature, and informatics*. Chicago: University of Chicago Press.

Heims, S. J. (1991). *Constructing a social science for post-war America: The Cybernetics Group*. Cambridge, MA: The MIT Press.

Henrich, J. (2020). *The weirdest people in the world: How the West became psychologically peculiar and particularly prosperous*. New York: Farrar, Straus and Giroux.

Kassarjian, S. (2021, July 14). Private Communication.

Kellogg, W. W., & Mead, M. (1976). *The atmosphere: Endangered and endangering*. Washington, DC: US Government Printing Office.

Kline, R. R. (2015). *The cybernetics moment: Or why we call our age the information age*. Baltimore, MD: Johns Hopkins University Press.

Kress, W. J. & Stine, J. K. (Eds.). (2017). *Living in the Anthropocene: Earth in the Age of Humans*. Washington, DC: Smithsonian Books.

Lewis, S. L., & Maslin, M. A. (2018) *The human planet: How we created the Anthropocene*. New Haven, CT: Yale University Press.

Marks, R. B. (2007). *The origins of the modern world: A global and ecological narrative From the Fifteenth to the Twenty-First Century*. New York: Roman & Littlefield.

Matthiessen, F. O. (1948). *From the heart of Europe*. New York: Oxford University Press.

McNeill, J. R., & Engelke, P. (2014). *The great acceleration: An environmental history of the Anthropocene since 1945*. Cambridge, MA: Belknap Press of Harvard University Press.

Mead, M. (1947). *The Salzburg Seminar in American Civilization* [Unpublished Report]. Cambridge, MA: Harvard Student Council. https://www.salzburgglobal.org/fileadmin/ user_upload/Documents/General_SGS_Documents/1947_MeadArticle.pdf

Mead, M. (1968). Cybernetics of cybernetics. In H. von Foerster, J. D. White, L. J. Peterson, & J. K. Russell (Eds.), *Purposive systems: Proceedings of the First Annual Symposium of the American Society for Cybernetics* (pp. 1–11). New York: Spartan Books. Retrieved October 26, 2021 from https://cepa.info/2634.

Meadows, D. H., Randers, J., Meadows, D., & Behrens, W. (1972). *The Limits to growth: A report for the Club of Rome's project on the predicament of mankind*. Washington, DC: Universe Books.

Osterhammel, O., & Peterssen, N. P. (2003). *Globalization: A short history*. Princeton, NJ: Princeton University Press.

Overton, B. J. (1995). Private conversation. (Salzburg Seminar, April 8, 1995).

Oxfam. (2021). *What is global citizenship?* Retrieved on September 1, 2021. https://www.oxfam.org.uk/education/who-we-are/what-is-global-citizenship/.

Pais, C. (Ed.). (2016). *Cybernetics: The Macy Conferences 1946-1953—The complete transactions*. Diaphanes.

Pickering, A. (2011). *The cybernetic brain: Sketches of another future*. Chicago: University of Chicago Press.

Prossnitz, G., Ryback, T. W., & Stegen, I. (1994). *Schloss Leopoldskron*. Salzburg, Austria: Otto Müller Verlag.

Raworth, K. (2017). *Doughnut economics: Seven ways to think like a 21st century economist*. White River Junction, VT: Chelsea Green Publishing.

Reade, C., Reckmeyer, W., Cabot, M., Jaehne, D., & Novak, M. (2013). Educating global citizens for the 21st century—The SJSU Salzburg Program. *Journal of Corporate Citizenship, 49*(March), 100–114.

Reckmeyer, W. J. (1989). *Strategic issues management: A renaissance systems perspective*. Menlo Park, CA: Electric Power Research Institute.

Reckmeyer, W. J. (1991). Managing complexity in the systems age: A renaissance systems perspective. In S. A. Umpleby & V. N. Sadovsky (Eds.), *A science of goal formulation: American and Soviet discussions of cybernetics and systems theory* (pp. 177–194). New York: Hemisphere.

Reckmeyer, W. J. (Ed.). (1993). *Leadership Readings: 1993-1994*. Stanford, CA: American Leadership Forum.

Reckmeyer, W. J., & Reckmeyer, W. J. [Père]. (1993). *Revitalizing America: Developing a coherent national strategy for the 21st century*. Reston, VA: Inter-National Research Institute.

Reckmeyer, W. J. (2005). The nature and use of systems-of-systems approaches in public policy-making and program management [Plenary presentation]. American Society for Cybernetics, *2005 Annual Conference*, Washington, DC, October 29, 2005.

Reckmeyer, W. J. (2010). *SJSU Salzburg Program—The First Five Years*. San José, CA: San José State University, International & Extended Studies.

Reckmeyer, W. J. (2011). *Global citizenship: US national strategy in a complex world*. Report on the 1st Annual Provost's Honors Seminar. San José, CA: San José State University, Office of the Provost.

Reckmeyer, W. J. (2013). *Educating global citizens*. TEDx Talk, San José State University. Retrieved October 26, 2021 from https://www.youtube.com/watch?v=TdEwSXFniIg

Reckmeyer, W. J. (2016a). Reflections on constructing a reality: The American Society for Cybernetics in the 1980s. *Cybernetics and Human Knowing 23*(1), 28–41.

Reckmeyer, W. J. (2016b). *Reflections on cybernetic praxis: Systemic actions to address complex challenges* [Norbert Wiener Gold Medal Address]. American Society for Cybernetics, 2016 Annual Conference, Olympia, WA, June 3, 2016.

Reckmeyer, W. J. (2019). *Homo cyberneticus: History of the Anthropocene.* Humanities Honors Program, San José State University, San José, CA November 5, 2019.

Reckmeyer, W. J. (2020). *Homo cyberneticus*: Future of the Anthropocene. Humanities Honors Program, San José State University, San José, CA April 4, 2020.

Reckmeyer, W. J. (2021). *Homo cyberneticus*: Creating and managing the Anthropocene. Club of Remy [Zoom presentation] June 16, 2021.

Rittel, H. W. J., & Webber, M. M. (1973). Dilemmas in a general theory of planning. *Policy Sciences, 4*, 155–169.

Rockstrom, J., & Klum, M. (2015). *Big world – small planet: Abundance within planetary boundaries.* New Haven, CT: Yale University Press.

Russon, C. & Ryback, T. (2003). Margaret Mead's evaluation of the first Salzburg Seminar. *American Journal of Evaluation, 24*(1), 97–114.

Ryback, T. W. (1997). *The Salzburg Seminar: The first fifty years.* Salzburg, Austria: Salzburg Seminar in American Studies.

Ryback, T. W. (2021). *The Salzburg Seminar—A community of fellows.* Salzburg, Austria: Salzburg Global Seminar. Retrieved July 15, 2021 from https://www.salzburgglobal.org/ about/history/articles/a-community-of-fellows.

Sachs, J. D. (2020). *The ages of globalization.* New York: Columbia University Press.

Salzburg Global Seminar (2020). *The Salzburg Global American Studies Program | History.* Retrieved July 15, 2021 from https://www.salzburgglobal.org/multi-year-series/american-studies/pageId/8856.

Salzburg Global Seminar (2021). *Salzburg Global mourns the loss of Mary Catherine Bateson.* Retrieved October 26, 2021 from https://www.salzburgglobal.org/news/latest-news/article/salzburg-global-mourns-the-loss-of-mary-catherine-bateson.

Schattle, H. (2008). *The practices of global citizenship.* Lanham, MD: Rowman & Littlefield.

Schudel, M. (2021, January 15). Mary Catherine Bateson, anthropologist and author of Composing a Life, dies at 81. Washington Post. Retrieved January 16, 2021 from https://www.washingtonpost.com/local/obituaries/mary-catherine-bateson-anthropologist-and-author-of-composing-a-life-dies-at-81/2021/01/15/1079431e-5771-11eb-a08b-f1381ef3d207_story.html.

Schwab, K. (2008). Global corporate citizenship: Working with governments and civil society. *Foreign Affairs, 87*(January-February), 108–118. Retrieved October 26, 2021 from http://www.foreignaffairs.com/articles/63051/klaus-schwab/global-corporate-citizenship.

Schwab, K. (2016). *The fourth industrial revolution.* Geneva: World Economic Forum.

Smil, V. (2012). *Global catastrophes and trends: The next fifty years.* Cambridge, MA: The MIT Press.

Smil, V. (2021). *Grand transitions: How the modern world was made.* Oxford, UK: Oxford University Press.

Smith, H. N. (1949). The Salzburg seminar. *American Quarterly, 1*(Spring), 30–37. https://www.salzburgglobal.org/about/history/articles/american-quarterly-article-1949.

Stearns, P. (2009). *Educating global citizens in colleges and universities.* New York: Routledge.

Steffen, W., Broadgate, W., Deutsch, L., Gaffney, O., & Ludwig, C. (2015). The trajectory of the Anthropocene: The great acceleration. *The Anthropocene Review* (April), 1–18. DOI: 10.1177/2053019614564785

Taleb, N. N. (2010). *The black swan: The Impact of the highly improbable.* New York: Random House.

UNESCO (2021). Global citizenship education. Retrieved September 1, 2021 from https://en.unesco.org/themes/gced

Von Foerster, H., Mead, M., &Teuber, H. L. (1950-1956). *Cybernetics: Circular causal and feedback mechanisms in biological and social systems.* New York: Josiah Macy Jr. Foundation.

Von Foerster, H. (1972). Responsibilities of competence. In *Understanding understanding: Essays on cybernetics and cognition* (pp. 191–197). New York: Springer Verlag.

Wright, B. F. (1948). Seminar in Salzburg. *Harvard Alumni Bulletin* (pp. 1–2).

Wucker, M. (2016). *The gray rhino: How to recognize and act on the obvious dangers we ignore.* New York: St. Martin's Press.

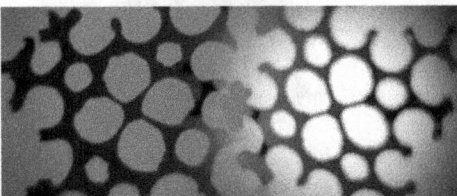

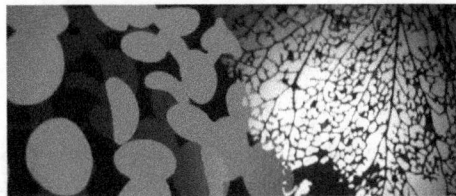

Lorusso, Mick. (2021). *Coherence.* Quantum Biology Series. Shadow performance video stills.

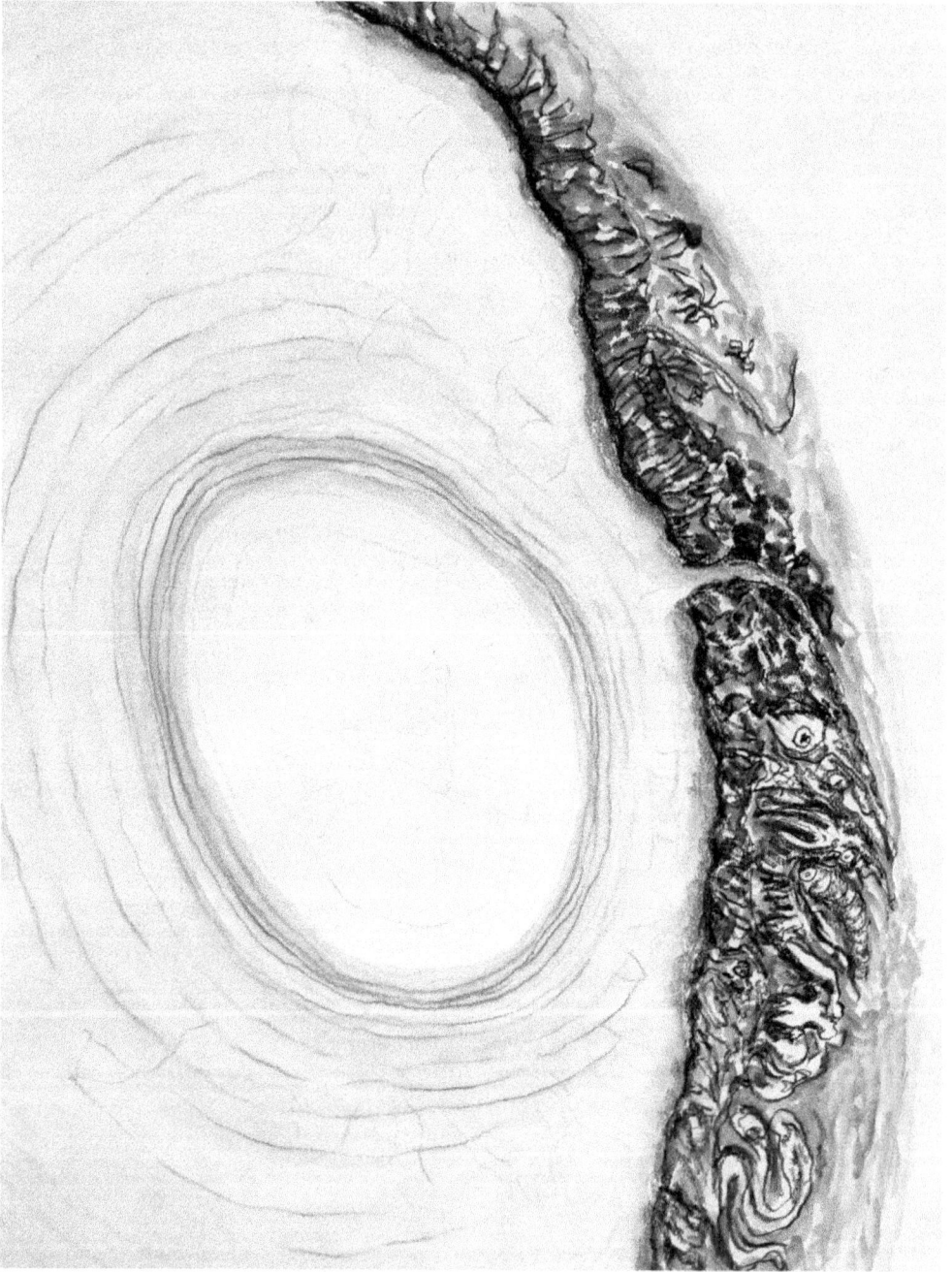

Lorusso, Mick. (2010). *The Crust Dissolves* (rotated.) Essence of Light/Life Series, Energy Patterns.
Drawing. Watercolor pencil on paper. 22 x 30 cm.

Cybernetics and Human Knowing. Vol. 28 (2021), nos. 3-4, pp. 103–110

Remembering a Message
From Mary Catherine

Jude Lombardi[1] and Larry Richards[2]

In August 2011, the American Society for Cybernetics recognized the many contributions of Mary Catherine Bateson to the field of cybernetics by awarding her the Norbert Wiener Medal. Mary Catherine could not attend the meeting for health reasons, though she did send a letter accepting the award and offering a warning and a challenge to the cybernetics community. The authors were both there and now, ten years later, reflect on that letter and its message.
Keywords: Whole systems, interconnectedness, violations of communication, conversation, participation, performance, interactor.

Introduction

We, the authors, unable to get together in person to discuss the letter that Mary Catherine Bateson sent to the American Society for Cybernetics (ASC) in 2011, decided to exchange questions, comments, interpretations, extensions and other reflections via emails, with occasional phone calls to help clarify our thoughts. What follows is an edited version of that exchange. Suffice it to say that Mary Catherine's letter provided much more material for us to explore than we had originally anticipated. Jude's initiating thoughts are presented first, followed by the letter and then our reflections. Because we mention a few well-known cyberneticians in our exchange, we include a final section with some notes on each and our personal connections to them.

Jude Lombardi: Mary Catherine was a lifelong contributor to cybernetic ideas and, for many decades, a presenter at the **American Society for Cybernetics (ASC)** conferences. For me, she was influential in the ASC shifting its focus away from objects, goals, purpose and papers toward rhythms, processes, performance and conversation. Her involvement in systems thinking and cybernetics, particularly her involvement in the **ASC**, were part of her cultural upbringing since her father, Gregory Bateson, and mother, Margaret Mead, were both key participants in the development of first and second cybernetics.

> First cybernetics is about... control, communication, goals and purpose.
> Second cybernetics is about... autonomy, conversation, processes and presence.

I first met Mary Catherine Bateson in 1993 at the ASC conference in Philadelphia, Pennsylvania. Over the decades I had the opportunity to interact with Mary Catherine

1. Video-ethnographer, cybernetician, bee steward. Baltimore, Maryland, USA, jlombardi@jlombardi.net.
2. Dialectician, organization designer, conversationalist. Portland, Maine, USA, laudrich@iue.edu.

on many occasions. Our paths often crossed at ASC meetings where we would share a meal and catch-up. She was always willing to share her calm, pausing before speaking, presentation of self. I will miss her presence.

It is an honor to coauthor this report about Mary Catherine with my dear friend and colleague Larry Richards. I am reminded of another cybernetic colleague **Annetta Pedretti** who once said it's not about answers—it's about what the next question(s) will be. So, I will reminisce about Mary Catherine with Larry Richards by asking him several questions that arise for me when thinking and writing about Mary Catherine and her messages to the ASC, and cybernetic communities worldwide.

In this paper, I will focus on her written message in 2011, to the ASC membership in response to receiving the Norbert Weiner Medal for her outstanding contributions to the field of cybernetics. In her letter she emphatically suggests possible necessities for meeting the need for a radical shift in our thinking (epistemology) and doing (performance) if we are not only to thrive but survive as species on the earth that we—and the other living creatures—are dependent on. After all, as she says in her letter, who else is going to do it?

Letter from Mary Catherine Bateson, ASC 2011

To my colleagues at the American Society for Cybernetics.

I write to express my regret at not being present to receive the Norbert Wiener medal with which you are honoring me, to express my gratitude, and to convey briefly what I would be saying if I were present, emphasizing the contribution that I believe the ASC can and should make.

Many of you will have read the essay by my father, Gregory Bateson, called "From Versailles to Cybernetics," in which he traces much of the madness of the 20th century, still ongoing, to violations of communication. He ends by declaring that there is "...latent within cybernetics the means of achieving a new and perhaps more human outlook, a means of changing our philosophy of control and a means of seeing our own follies in wider perspective." This hope rests on the potential offered by cybernetics for thinking in terms of whole systems rather than in terms of separate and competing interests and specializations, a potential that must be explored and expressed.

We are at a time of great danger, when the planetary cycles on which life depends and the long term patterns of climate are being severely disrupted. Meeting this danger and the humanitarian disasters that lie ahead requires a whole new order of cooperation. Yet researchers in the earth systems sciences have limited understanding of social systems, while some politicians deny what is happening, and non-specialists around the world simply do not recognize the larger picture. One day of cool weather leads to comments like, "See, the climate isn't changing after all." At the same time, the danger is amplified by an ideology that idealizes competition and accepts deception as a means to winning. Human beings do not always behave well when they believe that their "share of the pie" may be reduced, and modern weapons can turn the habit of zero-sum thinking into a lose-lose outcome for the entire planet.

Most of us understand this, but we need to remember how rare it is to participate in an intellectual community like this one, in which, for example, the acidity of the oceans, the instability of financial institutions, the rise of fundamentalism, and the increase in diabetes can be seen as examples of similar processes—and as possibly coupled. Most of us work within the framework of academic conventions that constrain scientists and scholars to keep such questions separate. Do we understand that in achieving new kinds of control we must bring all of our knowledge about communication and decision making to bear? Who else is going to do it?

I applaud your experiments with new formats for integrative discussion at this conference. It may be that the intellectual structure of cybernetics requires a new kind of communication that will make a new kind of listening possible, listening that carries the awareness of being part of a larger whole. If so, it must go beyond this small community. My hope is that all of us will resolve to carry our study of systems and cybernetics into our engagement with society, speaking out and strengthening exchanges with other fields and with the public, learning to think and then act to achieve the shared understanding and shared willingness to change so urgently needed. We need to be vocal and political. Somehow we must transform our shared understandings into a new kind of common sense.

Reflections

Jude: Larry, In the context of Mary Catherine's letter to the ASC, accepting the ASC's Norbert Wiener Medal in 2011, how is a focus on conversation relevant?

Larry Richards: I didn't get many chances to talk with Mary Catherine, although I did observe her presence and interaction with others at ASC meetings. Ranulph Glanville, then President of the ASC, did ask if I would read her letter at the 2011 ASC conference. Mary Catherine could not attend in person but wanted to express her appreciation for the award and take advantage of the opportunity to issue a warning to cyberneticians and systems theorists. It was my honor to read the letter.

Mary Catherine knew Gordon Pask; he was one of the attendees at the meeting on "Effects of Conscious Purpose on Human Adaptation" that she chronicled in her book, *Our Own Metaphor* (Bateson, 1991). While she may not have studied Gordon's conversation theory, she definitely interacted with others as though conversation was distinct from the common conception of communication. In conversation, there is a back and forth between two or more participants, each seeking to understand the other, explore their disagreements and possibly create new perspectives that none, by themselves, would have come to. In the common conception of communication, participants assume they already understand each other and agree on desired outcomes of their communication. In the cybernetic version of conversation, participants accept their differences (asynchronicities) as opportunities for creating the new (moving toward synchronicity). The value of conversation is as much, if not more, in the interaction itself as in any possible outcomes. Mary Catherine clearly placed high value on this form of interaction, regarding it as essential for addressing the complex, interconnected issues that we face as humans.

Jude: So, when in conversation, the autonomy of any respondent renders participants unpredictable in their responses and is relevant for any interactor.

I remember in her letter, Mary Catherine emphasized our need for remembering how rare it is to participate in an intellectual community like the ASC, a society in which "the acidity of the oceans, the instability of financial institutions, the rise of fundamentalism, and the increase in diabetes can be seen as examples of similar processes—and as possibly coupled."

When Mary Catherine talks about the existential issues facing humankind as possibly coupled, why does she suggest that cybernetics and systems theory might be useful, even necessary, for addressing the *whole*?

Larry: Mary Catherine would often speak of the interconnectedness of the various issues that threaten the well-being, if not survival, of humanity—issues like global warming/climate change, extreme inequality, terrorism and war, global health issues (e.g., diabetes, pandemics), environmental degradation, and so forth. When she says these issues are coupled, she is invoking a cybernetic idea. Early in modern cybernetics (1943-present), the idea that systems treated as separate could be connected (coupled) to other systems through a causal relation between single variables in each system presented a problem for traditional scientific thinking. In the biological and social sciences, the conditions of separability, assumed in the physical sciences, did not necessarily apply. Either the systems had to be expanded to try to encompass the whole, or the conditions under which separability could be reasonably assumed had to be identified and adhered to when drawing conclusions. This led to the idea of *whole systems theory*, where every system is assumed to be incorporated in larger systems, and the larger the systems we formulate and model, the more likely we are to address what otherwise would be unanticipated consequences. However, this relational and wholistic approach to coupling is not the only way of thinking about the interconnectedness of all things. Issues and systems can be dynamically connected. That is, a change in one system can potentially trigger a change in another system. This is not a causal connection, as the change does not result in a specific, hypothetically predictable change in the other system; it rather perturbs the other system in a way that taps its potential to respond, but not in a predictable way, even hypothetically. When we say that living (autonomous) systems are structurally coupled, we are saying that there is a pattern of dynamics in the interactions between/among them. This is also a way of thinking about the human phenomenon of conversation, as opposed to communication.

Jude: In this context, communication is causal and conversation dynamic. So, in conversation the focus is on processes and presence rather than purpose and progress.

In her acceptance letter, Mary Catherine reminds us of her father's essay, "From Versailles to Cybernetics," writing about how he spoke of the problems and disasters we faced then and, as she points out, now (population, climate and the occidental ideas of man). Echoing her father's essay, she suggests that many of these problems stem from our violation of communication.[3]

What is the "violation of communication" discussed by Gregory Bateson in his paper "From Versailles to Cybernetics"?

Larry: In "From Versailles to Cybernetics," Gregory Bateson (2000) identified what were to him the two most historically important events to have occurred in the 20th century at the time he gave the lecture (1966)—the circumstances surrounding the preparation of the Treaty of Versailles (1921) and the Josiah Macy, Jr., Foundation meetings on cybernetics (1946-1953). Both events, he claimed, changed the underlying epistemological basis for acting in the world—the former leading to

---

3. Mary Catherine suggests a need for a new kind of communication, one nested in forms of listening that include an awareness of each one's participation in all that we do.

potentially disastrous results for the world and the latter offering a new way of thinking about possibilities for a more sustainable world. The preparation of the Treaty of Versailles, bringing an end to World War I, introduced deception as an acceptable form of behavior in negotiations among adversarial nation-states. Bateson traces the circumstances that led to World War II to this Treaty, as well as to a way of thinking that the most acceptable ending to a war is complete annihilation of the enemy (which we saw at the end of WWII in both Germany and Japan). Whereas communication that would lead to a treaty had previously assumed that the participants could rely on statements made and offers tendered, the reversal of such statements and offers that characterized the proceedings surrounding the Treaty of Versailles represented a violation of the understood rules of communication.

Jude: So, in this context, communication is violence. Herbert Brün said: "Insistence on communication ultimately leads to social and physical violence" (Brün, 2004, p. 289).

Our situation in Afghanistan reflects our still being stuck in communicating rather than conversing.

Larry: ... and still relying on deception to win at all costs, which Mary Catherine referred to as an example of a common sense idea that is problematic.

Jude: What about the other event to which Gregory refers, the Macy conferences on cybernetics?

Larry: The Macy Conferences on cybernetics, on the other hand, offered an alternative epistemology, a way of thinking and knowing and a set of values that recognized that world problems are interconnected and that those interconnections embed circularities and processes whose complexity surpasses our individual capabilities to address them. That is, there are ways of acting in the world that recognize, as Mary Catherine points out in her letter, that we humans and our systems are part of wholes much greater than ourselves, complexities that our traditional ways of thinking cannot adequately address. The cybernetic alternative may not give us the certainty that we have come to think is possible and to demand—that is, there are no guarantees; however, the alternative may provide humanity with its best chance for survival, at least for a longer period of time.

Jude: I remember Humberto Maturana questioning the use of the term *wholeness* (holism) as relevant when thinking cybernetically.

How do you think Mary Catherine would bring together the cybernetic ideas of interconnectedness, conversation, and participation?

Larry: When Mary Catherine talked about systems being a part of a larger whole, I don't think she was suggesting that we should expand the models we use to address world issues until they include everything possible. This is a hierarchical approach to the problem—systems within systems, within systems, and so forth. And, systems that include everything distinguish nothing. Rather, I think she was pointing to the fact that we often erroneously assume we can know enough to solve these interconnected problems, as though we are outside of them. In other words, we need to think, individually and collectively, that we must include ourselves in our formulations of the systems we are addressing and recognize that there is always a larger whole that we cannot fully comprehend. So, an alternative to thinking in terms of whole systems as systems within systems is to think in terms of the patterns of dynamics in the interactions among systems, among observers, and between observers and their systems—that is, in terms of processes. If our actions are to be in the interests of humanity, these processes must be participative. The more perspectives and insights that can be brought to bear on the issues of concern, the more likely are approaches to those issues to be robust. Participation implies that people are aware that they make a difference, even if they do not know and cannot trace the difference made on specific outcomes. Cybernetics provides a way of thinking that supports this assertion, with conversation being the process through which participation can occur. This view of participation suggests that people do not need to be in positions of power and influence to make a difference. They can make a difference through the dynamics of interaction generated through a network of conversations. The cybernetic epistemology that supports this view is a participative and emancipatory one. Rather than thinking in terms of specific goals, think in terms of desirable processes. Participating in desirable processes is more likely to produce desirable outcomes than specifying the outcomes first and then trying to achieve them. There is, of course, much more to think and talk about: that's the process.

Jude: Yes, processes, rhythms, and presence (Pedretti, 1993).

When you speak of desires it reminds me of the Desires Exercise as described in Manni Brün's little book *Designing Society* (1985).

One last question for now: What about performance?

Larry: The composer and cybernetician Herbert Brün spoke of performance as: "Sharing your presence; conveying your thought and your intention; carrying your messages so that they reach out the way you want" (Brün, 2003, #118, p. 6). So, performing is acting with intent, being in the present, with thought and awareness of our desires and possible consequences. Performance is not about pretending to be someone I am not; to the contrary, it is about becoming a thinking, caring person who takes responsibility for my actions. When Mary Catherine spoke about a needed shift in our thinking and acting, I speculate that she may have been speaking of turning our actions, including our behaviors in conversations, into performances. An important role for performance in society is to provoke conversation. An action undertaken as a performance not only has consequences for the target of the action but also has consequences for the conversations it provokes. It does the latter through the particular dynamics of the performance, not only through the causal intentions of the actor. I contend that paying attention to the dynamics of our actions and (especially) interactions is a derivative of cybernetic thinking. When speaking, paying attention to dynamics involves treating rhythm, speed, volume, pitch, emphasis, pivots, pauses, and so forth, with intent. I would also argue for the cybernetic version of intention as an awareness of our desires as constraints on the processes (and actions) in which we engage (in the present), rather than as goals or objectives to be achieved (in the future). This is, obviously, more than a simple shift in thinking; it is a fundamental shift in our language and the way we interact with each other. Perhaps, this is why Mary Catherine targeted cybernetics and systems theory for her remarks.

Jude: I would say it requires a fundamental shift in our languaging which entails all that we say and do. Alas a provocation for another time.

Thanks Larry, for this almost conversation.

May Mary Catherine Bateson live on in languaging for decades to come.

Bee well.

Larry: Thank you, Jude. It is always a pleasure, and my best to your bees—we need them!

Some Notes on Cyberneticians Mentioned

- Gregory Bateson was among the original participants in the Macy meetings on cybernetics in the 1940s and 1950s. His contributions to cybernetics, systems thinking, mind, culture, power, psychosis, and research methods have influenced many of us.
- Margaret Mead, well-known as an anthropologist, was also present at the Macy meetings on cybernetics. Heinz von Foerster credits her with instigating what he would call second-order cybernetics through her paper, "The Cybernetics of Cybernetics" (1968). Mary Catherine wrote about her parents in *With a Daughter's Eye* (1984).
- Annetta Pedretti was a student of Gordon Pask at the Architectural Association in London. Language was a central theme of her work and writings. Her unique approach to publishing (Princelet Editions) and participating in ASC conferences produced insightful interactions for both of us.
- Ranulph Glanville, also a student of Gordon Pask at the Architectural Association, served as President of the ASC for six years until his death in December 2014. He is often credited with connecting, in a deliberate way, design and the design community to cybernetics and the cybernetics community.
- Gordon Pask is especially well-known for his development of conversation theory. Through his influence, many now consider cybernetics to be enacted in conversation. Larry regards him as one of his mentors in cybernetics and conversation, along with Herbert Brün and his first teacher and mentor in cybernetics, Klaus Krippendorff.
- Herbert Brün, composer, graphic artist, pioneer with computers in art and music, participated in Heinz von Foerster's Biological Computer Lab and Heinz's course on *Cybernetics of Cybernetics* (1995) at the University of Illinois. He was a member of Jude's doctoral committee and mentor in cybernetics for both of us over many years.
- Humberto Maturana, biologist and cognitive scientist, also spent time at the Biological Computer Lab at the University of Illinois. It was during this time that he, Francisco Varela, and Ricardo Uribe developed their seminal concept of autopoiesis. He was also a member Jude's doctoral committee. His ideas on the

biology of language, languaging, cognition, and love have influenced us profoundly as well as many others in and out of cybernetics.

•   Marianne Brün taught a course at the University of Illinois that morphed into the Princelet Editions book, *Designing Society*. The course became the inspiration for the School for Designing a Society, that began in 1992 and continues in various forms to this day. Both of us have participated often in this school, and the impact of Manni's presence cannot be overestimated.

References

Bateson, G. (2000). From Versailles to cybernetics. In *Steps to an ecology of mind* (pp. 477–485). Chicago: University of Chicago Press. (Originally published in 1972.)

Bateson, M. C. (1991). *Our own metaphor.* Washington, DC: Smithsonian Institution Press. (Originally published in 1972.)

Bateson, M. C. (1984). *With a daughter's eye: A memoir of Margaret Mead and Gregory Bateson.* New York: HarperPerennial.

Brün, H. (2003). *Irresistible observations.* Champaign, IL: Non Sequitur Press.

Brün, H. (2004). *When music resists meaning* (A. Chandra, Ed.). Middletown, CT: Wesleyan University Press.

Brün, M. and respondents. (1985). *Designing society.* London: Princelet Editions.

Foerster, H. von (Ed.). (1995). *Cybernetics of cybernetics.* Minneapolis, MN: Future Systems. (Originally published in 1974.)

Mead, M. (1968). The cybernetics of cybernetics. In H. von Foerster, J. D. White, I. J. Peterson, & J. K. Russell (Eds.), *Purposive Systems* (pp. 1–11). New York: Spartan Books.

Pedretti, A. (1993). Turning objects into rhythms. Paper presented at the American Society for Cybernetics Conference, November 3-7, 1993 in Philadelphia, Pennsylvania. Accessed September 14, 2021 at: https://www.youtube.com/watch?v=6Q78GHYle0s (start at: 4:30 min.)

Lorusso, Mick. (2006-08). *Tree Being.* Anima Mundi Series, Energy Patterns.
Drawing. Graphite on paper. 45 x 60 cm.

Cybernetics and Human Knowing. Vol. 28 (2021), nos. 3-4, pp. 111–120

Braiding Continuity and Improvisation
Let's Go Exploring!

Frederick Steier[1]

This article builds on Mary Catherine Bateson's work on the construction of continuity and its cybernetic basis. The relationship between continuity and discontinuity in how we envision our life paths is developed, and then extended to other systemic domains. Based on an exercise that Bateson crafted involving two paths for creating life narratives, themes arising from the juxtaposition of continuity and discontinuity are presented. Reflections by the author on what emerged with exercises with co-learners based on Bateson's ideas are offered together with extensions of her themes. The importance of flexibility and improvisation as key emergent principles is discussed, as are aspects of what opportunities and challenges are afforded when continuity and improvisation are braided together. Extensions to other domains of learning, including environmental learning are presented as featuring ideas of an improvisational excellence. An invitation for playful exploring is made.

Keywords: improvisation, flexibility, continuity, discontinuity, cybernetic learning

A text that I continually return to for inspiration is the final cartoon strip of Bill Watterson's *Calvin and Hobbes* (Watterson, 1995/2005). It was originally published on December 31, 1995, a New Year's Eve offering. It was known that it was to be the final strip, offering thoughts for the new year—and beyond—as a farewell. It is a three-panel strip, with two of the panels having panels embedded within a larger panel. In the first panel, Calvin is walking in front of Hobbes, who is carrying a toboggan, in a snow-covered landscape. Calvin remarks, "Wow, it really snowed last night! Isn't it wonderful?" Both Calvin and Hobbes are smiling broadly. Embedded within this panel is another panel, with Hobbes saying to Calvin "Everything familiar has disappeared! The world looks brand new!" while Calvin tells Hobbes "A new year … A fresh clean start!" The following panel finds Hobbes, still smiling broadly, saying "It's like having a big white sheet of paper to draw on!" while Calvin, looking thoughtful and curious, continues with "a day full of possibilities." In the following and final panel, on the left we have the embedded panel, with Calvin turning to say to Hobbes, "It's a magical world, Hobbes, Ol' Buddy" with Hobbes and Calvin boarding the toboggan. In the larger frame of that final panel, we see Calvin and Hobbes both grinning widely, sledding down the snow-covered hill, with Calvin continuing, eagerly "… let's go exploring!" I am smiling as I write this, such are the memories and thoughts for the future that this strip evokes for me.

Of course, as with most strips, knowing the history helps to appreciate what is being offered. Calvin is a young boy, and Hobbes his pet toy tiger who is animate in Calvin's presence, but not for anyone else. They have shared poignant moments together over the more than ten years of the strip, while also, throughout the life of the

---

1. Fielding Graduate University and University of South Florida. Email: fsteier@gmail.com

strip, commenting on codes and modes of everyday life, with each other and with others, including Calvin's parents. The final strip, among many other things, juxtaposes a way of starting anew, to go exploring on the blank page of the snow-covered landscape, while holding to the continuity of their exploration and curiosity-soaked relationship. It is simultaneously an affirmation of tradition and a new beginning, intertwined as an invitation for readers from the final publication of the strip.

Constructing Continuity

The very idea of starting anew while continuing the old, of the connection between continuities and discontinuities, is a major theme of cybernetics and systems. It shows up in various forms, for example in questions about how we balance tradition and change or identity and transformation, such as we see in Gregory Bateson's *Mind and Nature* (G. Bateson, 1979). It also shows up in Ross Ashby's development of the central idea of ultrastability in his *Design for a Brain*, where keeping essential variables within limits while adapting to environmental change becomes a focus (Ashby, 1952). And it is a theme developed in different ways throughout the career of Mary Catherine Bateson, which is the core concern of this article.

In "The Construction of Continuity" (Bateson, 1992), Mary Catherine Bateson develops a framework for understanding how we construct and experience the idea of continuity by inviting us to consider ways of bringing together continuity and discontinuity. Even though Bateson takes as her starting point how we construct our lives and life stories, she also offers implications for organizational and institutional continuity and discontinuity. Indeed, the bringing together of continuity and discontinuity in the construction of life narratives is also central to Bateson's *Composing a Life* (Bateson, 1989), on which the 1992 essay builds, and to her subsequent, *Composing a Further Life* (2010). In *Composing a Life*, Bateson explores the lives of four women in terms of how each crafted paths toward meaningful and successful lives, while responding to challenges, and discontinuities along the way. Bringing her own life story into the book as well, Bateson notes that the very idea of responding to discontinuity in a career path, is especially prominent for women although it can of course be the case for men. The stories offer compelling illustrations, as themes of flexibility, comfort with ambiguity, and responsiveness to new ways of dealing with novel situations, emerge. Central to all the stories are cybernetic ideas concerning the importance of valuing non-linearity.

I am intentionally choosing the essay, "The Construction of Continuity," as a starting point since it extends ideas from *Composing a Life*, yet ostensibly speaks to a different audience more in the realm of organizational life than much of Mary Catherine's previous work. At the same time, it is important to recognize that for those who had been following Mary Catherine's work across different scenes, it was also a continuity. The theme of continuity and discontinuity is reflected in the relationship between the essay and its diverse audiences.

Bateson begins this essay with a story of a workshop exercise she had given on "Emerging Ambitions," held as part of a computer conference. Carefully crafted to open up space for a bringing together of continuity and discontinuity and for exploring the tensions that this might engender, the workshop provides a prompt for Bateson's reflections that are the basis of her essay. At the workshop, Bateson invited the participants to offer two brief versions of their life story, with each tied to a different request for its telling. First, Bateson asked the participants:

> How would you tell the story of your life as an elaboration of the following statement: Everything I have ever done has been heading me for where I am today? (Bateson, 1992, p. 27)

And then,

> How would you tell the story of your life as an elaboration of this other statement: After lots of surprises and choices (and perhaps interruptions and disappointments), I have arrived somewhere I could never have anticipated? (Bateson, 1992, p. 28)

It is important to recognize the multi-faceted communicative valence of the ask, and its relational dimension, in the context in which the exercise is being performed. For example, the way Bateson poses the prompts invites the audience to engage in a novel performance while attending to the relational aspects of asking and hearing the questions. Rather than just asking for a report of preexisting facts, Bateson has invited the participants to invent different versions of what might or might not be the same story for them, and to do so in a relationship with an inquirer, Mary Catherine, who has positioned herself to learn with them, rather than just about them. The relational aspect of inquiring in a context has also been a theme throughout Bateson's career (see Bateson, 1991).

In her reflections on what ensued, Bateson expresses her own learning about continuity and discontinuity in themes familiar to those in the cybernetics and communication communities, while carefully using language that those in the technology and organization communities might also find resonant with their lives. Bateson is also careful to note that the theme of the exercise, "Emerging Ambitions," carries a bias toward discontinuity. In this way, the continuity version against the backdrop of the discontinuity version can take on a character of "yes, and," as compared to a "yes, but," with the "yes, and" finding attunement with those who value improvisation.

Emergent Themes of Continuity and Discontinuity

I turn now to three of the themes that emerged for Bateson from bringing continuity and discontinuity together. Recognizing the cybernetic and systemic aspect of the exercise, I will also explore patterns across the themes and try to bring these themes into the present, by connecting them to some of my own experiences using this exercise.

One major theme that emerged for Bateson centers on a capacity to transfer learning. As Bateson notes, "If a situation is construed as totally new and different, earlier learning may be seen as irrelevant. The transfer of learning relies on some recognized element of continuity" (Bateson, 1992, p. 32). In other words, the participants recognized that in order to learn from one situation to the next, it is necessary to see what might be similar in different situations. At the same time, we also recognize the importance of maintaining distinctions of context. Transfer of learning is one way of establishing continuity in our lives or in the lives of our organizations or communities. We can extend this a bit to also recognize, as Gregory Bateson did, the importance of levels of learning (G. Bateson, 1972). In other words, in bringing a lens of continuity to discontinuities, we create a context for appreciating learning to learn, or deuterolearning, as we recognize patterns of continuity. Establishing some continuity across seemingly disparate settings invites readiness for learning at multiple levels.

This connects with a second theme identified by Mary Catherine, which is an appreciation of logical levels. That is, our assumptions about what continuity is needed in order for other things to change may not be at the same logical level as that which we see as changing. Similarly, what we see as needing to change in order to maintain stability may not be at the same logical level as that which we want keep the same. Bateson notes that, "All change can be interpreted as an effort to maintain some constancy, such as survival" (Bateson, 1992, p. 36), a principle guiding programs in sustainability.

A familiar example is how in riding a bicycle, we seek to maintain our balance as well as our position relative to the ground—we would like to stay upright. Yet to do this, we need to have the flexibility to change many other things—such as how we lean into the wind or navigate a turn. If we are too rigid, we fall over. Similarly, in talking about an ecosystem's viability, we are speaking at a different logical level than if we talk about the processes of change that maintain that viability. Bateson insists on the need to appreciate differences in logical levels. In these examples we see the importance of maintaining flexible behaviors at the lower level as critical to how we construct continuities at a higher level. It's not that continuity and discontinuity exist at different levels, but that what we often use to frame what is continuous or to describe it, can be at a different level than what we see ourselves as needing to change. How we understand the role of flexibility in the juxtaposition of continuity and discontinuity emerges as significant, which leads into a third theme.

This further theme that Bateson identifies is the pathologies of continuity, which entails the idea that in holding on something to be maintained, we may also introduce a kind of stuckness. In other words, what we are holding on to may no longer be viable, or even desirable, in changing environments—or, on some occasions, in the same environment. This stuckness may even take the form of addiction—addiction to what we see as essential, including to ways of being. In developing her perspective on the construction of continuity, Bateson identifies two forms of a pathology of continuity. One form would be analogous to maintaining a constant level of intake or consumption. For this type of addiction/pathology, Bateson notes, "the relationship between levels is distorted: a superficial constancy is sustained at the cost of more profound change" (Bateson, 1992, p. 38). Bateson contrasts this with a second form, where "the need is not for a given substance but for a constant change in the supply of that substance, which must steadily increase because of various forms of habituation and changes in threshold" (p. 38). In other words, what the continuity rests on is an addiction to an increase, not just to a constant level. In ecological terms, we might wonder, as Bateson does, if an addiction to growth is an example of the latter pathology of continuity. Even so, Bateson notes that there are forms of the latter that may be positive—such as a desire for continuous learning, a point I return to later in the essay.

A Learning Exercise

These three themes identified by Bateson through her consideration of continuity and discontinuity as alternative frames, gives rise to further question. We might ask, for example, what the connection is between 1) an ability to transfer learning to new contexts while recognizing differences in context, 2) an appreciation of the relationship across logical levels of continuity and discontinuity, and 3) an awareness of potential pathologies of continuity and change. A question that emerges, in the context of our life narratives, which is also relevant for our communities and organizations, is what do we need to hold on to to maintain a valued identity, and how might we hold on to the variety necessary to adjust to future circumstances? In other words, how might we construct ways of bringing continuity and change together? To explore this a bit more, I draw examples from an exercise that I led based on Mary Catherine's workshop.

The occasion presented itself at a graduate university's new student orientation. As part of a residential session intended to both orient and build community among the new cohort of doctoral students, small groups of students were invited to create personal narratives centering on how they happened to arrive at where they are now. The program is tailored toward mid-career professionals seeking a doctorate, so the students were generally older and had a wider variety of experiences than what might be regarded as typical entering doctoral students. The new student orientation had a rich tradition that preceded my participation as a host, with the personal narrative as a central organizing feature. Being familiar with Mary Catherine's exercise, and sensing that it fit wonderfully within the context, I invited each of the participants in the small group I was working with to spend some time reflecting on two different paths for creating their personal narratives—one framed by discontinuity, and one framed by continuity. I invited them to also bring visual modes, such as drawing, into their telling and gave them large butcher block paper along with crayons and markers to aid them in creating two different visual depictions of the stories. My aim was to introduce a kind of playfulness into their exercise, while also recognizing that for some, the opportunity for serious self-reflection would loom large.

The exercise, in this context, turned out to be a powerful vehicle for participants' reflections about their life paths. More significant than devising the two paths for the stories was the juxtaposition of the two ways of telling. The group discussion was enlivened not only by the content of the stories themselves but by what came into view by bringing the two versions together. As Mary Catherine observed in performing the exercise with the computer group, the occasion itself may have afforded a bias towards a frame of change and discontinuity, given that the new student orientation was an occasion celebrating a movement across a boundary into a new program. As many of the participants were also mid-career in their professional lives, this boundary crossing was a key feature of the larger context, as many had been away from a university setting for some time.

Although a rich variety of themes emerged, each of the three themes identified above appeared prominently. To begin, while all participants delighted in telling the two different stories, I noticed on reflection that the process of juxtaposing two paths of storytelling had surfaced Mary Catherine's conceptual themes in new ways. Furthermore, having both versions—and offering them in a particular order—seemed to matter, as they sometimes reflected on how a similar event changed its shape with the second telling. Systems principles, in particular those of emergence, and the cybernetics of recursion were woven through the session.

Although I am offering illustrations of how each theme played out, the themes need to be seen as interrelated. For example, the recognition of the importance of the transfer of learning across contexts loomed large. One participant noted how she had studied art and actually worked for many years as a graphic artist, largely because she enjoyed forms of visual expression, but also because she had been identified at an early age as having artistic talent. At some point, as she described it, she made a left turn and trained for and decided to become an executive coach. In doing the exercise, she noted how much of the creativity, and reliance on visual forms of expression, often taking the form of bringing visual metaphors into her coaching relationships, was a carry-over from her art background. At the same time, she noted how she had not really thought about those continuities and connections in precisely that way. Smiling broadly, she told the two stories, which she also drew as different selves.

Another told how doing the exercise made her realize now that the thread that held her life together through various pivot points and discontinuities, was her love of learning. This included learning in different situations, as she even joked about what was continuous was precisely for her, the value of novelty. She expressed a lot of trepidation about beginning a doctoral program as she had been out of university life for many years, but could now joke about being in competition with her daughter, who was also a doctoral student, albeit a more traditional one. Her telling evoked the realization of the importance of logical levels, with love of learning being the thread that connected the various scenes for her. Interestingly, she chose the word traditional as a contrast to how she saw herself, as the language of change loomed large in the exercise.

One other participant told of dramatic shifts in his occupations over the years, while noting that what he had evolved to be doing had little relationship to what he had studied as an undergraduate and graduate student. He had studied architecture, but had been working in various tech start-ups, in very different arenas, since his earlier university days. In telling the story of continuity, however, he stated that he realized that he was addicted (his word) to challenges and saw each shift as a way to create a new challenge. He even reflected that he often felt unprepared for the content of the new arenas he had entered, but that was, for him, part of the challenge.

Another story that was offered centered on bringing material objects to create a continuity of place in response to discontinuities of location over time. Observing that we were in a conference center, this participant explained that it was important to her to bring a sense of at-homeness to wherever she was. Her professional life over the years required much travel and what stood out to her from her childhood in a military family was moving a lot. As an adult she resolved this by bringing a rolled up poster of a homey scene to wherever she was and hanging it in her hotel room. Her narratives of continuity and discontinuity both centered on her desire to holding on to a sense of place. The poster served to construct a kind of semiotics of continuity by sustaining her sense of at-homeness.

Each of these excerpts from larger narratives can be seen to connect with a particular theme, whether about continuity of learning, different levels of continuity and discontinuity, or a recognition of how continuity can become a stuckness—even if that stuckness is one that may be desirable. At the same time each excerpt illustrates the interconnectedness of the themes. And more importantly, is how the bringing together of the two paths of evoking a life narrative invited both reflection and seeing a background in a new way, while also offering ideas about anticipating and enacting futures. I also shared versions from my own life during the workshop as a way of creating a transition from the participants' stories to a general dialogue about what we learned collaboratively. Some snapshots follow.

My degree is in Social Systems Sciences from a program emphasizing systems and cybernetics, while also having a basis in action-oriented research. Over the course of my life I have, among other things, been a director of research at a child guidance clinic where systems approaches to family therapy thrived, as well as a director of a Center for Cybernetic Studies in Complex Systems in an engineering management program, where I also led a program on knowledge sharing and culture change at NASA. Other places have included being a visitor in an anthropology department as part of a project on overheating and sustainability, and a director of Interdisciplinary Studies Programs while also being a scientist-in-residence at a science center trying to bring systems and cybernetic principles to exhibit design. These seem like huge career changes (discontinuity looms large) but my continuity story brings into focus my fascination with cybernetics and systems. Each setting, was an opportunity to explore systems and cybernetic processes, including second-order processes, in diverse settings and to honor, in Gregory Bateson's words, the patterns that connect. Some of the patterns that connect centered on honoring key figures in the history of cybernetics, as opportunities to explore designs for cybernetic learning (Steier & Ostrenko, 2001; Steier & Jorgenson, 2003; Steier, 2005).

On the Road to Improvisation

My own stories, with the variety of cybernetic scenes, prompted a conversation within the group, as ideas operating at higher logical levels such as love of learning, desiring challenges or noticing cybernetic processes, loomed large. As the participants began to see interconnections within their own stories and between their own and each other's, they discovered the necessity of improvising in new situations while transferring learning from prior contexts. Having the requisite flexibility is essential to enact any improvisational performance. As we weave continuity and discontinuity in our lives and bring it into the lives of communities, families and other social or ecosystems, a new understanding of creativity emerges. It is precisely this kind of everyday creativity that is central to Mary Catherine Bateson's work. In her essay on ordinary creativity she invites us to value those situations where we do not have a ready script telling us what to do (Bateson, 1999). Often we are called to engage in a new performance in concert with others and in, what are for us, novel contexts. In recognizing this ability to respond to new situations in ways that also retain what we hold as dear, she makes clear that creativity of this sort is an important feature of everyday life, although it is rarely formally recognized as creativity. What Mary Catherine is referring to as ordinary creativity is often performed in relationships with others who may have different codes of behavior, and who may be operating in frames other than those we are accustomed to acting within. As such, ordinary creativity requires an ability to improvise and to have a flexible repertoire for acting. This flexible repertoire can include realizing that the very frames that we use to mark a situation as *that situation* must also be open to exploration, as they may not fit with the perspectives of others who are operating in concert with us. The very idea of frame flexibility (Steier, 2005) becomes central as an ability to maintain a variety of ways of being.

Earlier, I noted how those who make improvisation part of their performance emphasize the idea of a "yes, and" rather than "yes, but" (Leonard & Yorton, 2015) orientation. "Yes, and" is meant to recognize how any idea can be built on and responded to in ways to bring forth new possibilities. It can allow us to get unstuck. Groups who feature improvisation at the heart of their performances, particularly comedy performance groups, recognize the importance of "yes, and" as a way of valuing potential ambiguity. At the heart of "yes, and" can be an openness to recognizing that multiple interpretations of what is going on around us are both possible, and potentially central to another. To be able to improvise, we need to have a variety of possible understandings of any situation and to recognize that the frame we place around any situation may be different for others. My concern in this essay, and what I value throughout the work of Mary Catherine Bateson, is that an understanding of difference is central to how we bring continuity and discontinuity together.

Although improvisation and continuity might be seen as moving in different directions, there is an ecology of improvisation that sustainability requires.

Further, there is a tension between continuity and discontinuity on which improvisation rests. Much as in how dialogue requires holding a tension between contrasting views (Buber, 1970), improvising entails valuing that tension while bringing forth the unforeseen and unheard.

Improvisational Excellence in Environmental Learning

In this essay, I have focused mostly on continuity and improvisation in how we live and experience our lives and the lives of others. We can extend our exploration more outwardly to the lives of our environments and ecosystems. Mitchell Thomashow elegantly builds on Mary Catherine's work on the place of improvisation in crafting our lives, by extending it to environmental learning (Thomashow, 2020). In *To Know the World*, Thomashow links improvisation in environmental learning to themes that resonate deeply with second order cybernetics, as he invites us to walk in the forest with him, observing our observing, listening in new ways when walking, recognizing the music of the forest, and allowing ourselves to be in flow and in play with an environment. In particular, and central to improvisation, is how Thomashow requests that we honor the random in a multitude of ways, including the playfulness that the random can evoke (see also Nachmanovitch, 1990). Thomashow invites us to value improvisational excellence as key to environmental learning that does so in relation to self-learning. In short, to bring improvisation and learning together by exploring ourselves and our worlds.

As Mary Catherine notes in her classic article on joint performance across cultures, "My emphasis on improvisation and on learning going on all the time is a way of escaping from the notion that you are supposed to get something under your belt, some fixed body of knowledge: control it, have it, be educated and then live a life. If we can get away from that we may be able to reshape the current, very badly shaped, debate about education in such a way that we have some sense of the resources available for learning and of how life can keep on unfolding as an artistic performance, as a graceful and creative improvisation, for people of every age" (Bateson, 1993, p. 121).

Let's go exploring with Calvin and Hobbes! And bring Mary Catherine's ideas with us as we do.

Acknowledgments

The author would like to thank Jane Jorgenson for her helpful comments on earlier drafts. He would also recommend to readers Stephen Nachmanovitch's tribute to Gregory Bateson, "Old Men Ought to be Explorers," which invites us to walk similar paths.

References

Ashby, W. R. (1952). *Design for a brain*. London: Chapman and Hall.

Bateson, G. (1979). *Mind and nature: A necessary unity*. New York. Bantam Books.

Bateson, G. (1972). *Steps to an ecology of mind*. New York: Ballantine.

Bateson, M. C. (1989). *Composing a life*. New York: Atlantic Monthly Press

Bateson, M. C. (1991). Multiple kinds of knowledge: Societal decision-making. In M. J. McGee-Brown (Ed.), *Diversity and design: Studying culture and the individual* (pp. 1–21). Athens, GA: College of Education, The University of Georgia.

Bateson, M. C. (1992). The construction of continuity. In S. Srivastva, R. F. Fry & Associates (Eds.), *Executive and organizational continuity: Managing the paradoxes of stability and change* (pp. 25–39). San Francisco: Jossey-Bass.

Bateson, M. C. (1993). Joint performance across cultures: Improvisation in a Persian garden. *Text and Performance Quarterly, 13*, 113–121.

Bateson, M. C. (1999). Ordinary creativity. In A. Montuori & R. E. Purser (Eds.), *Social creativity, volume 1* (pp. 153–171). Creskill, NJ: Hampton Press.

Bateson, M. C. (2010). *Composing a further life: The age of active wisdom*. New York: Knopf.

Buber, M. (1970). *I and thou*. New York: Scribner.

Leonard, K. & Yorton. T. (2015). *Yes, and. How improvisation reverses "no, but" thinking and improves creativity and collaboration: Lessons from the Second City*. New York: Harper Collins.

Nachmanovitch. S. (1981). Gregory Bateson: Old men ought to be explorers. Retrieved February 8, 2022 from https://www.freeplay.com/Writings/Nachmanovitch.Bateson.Old.Men.Ought.To.Be.Explorers.d.pdf

Nachmanovitch. S. (1990). *Free play: Improvisation in life and art*. New York: Penguin Putnam.

Steier. F. & Ostrenko, W. (2000). Taking cybernetics seriously at a science center: Reflection-in-interaction and second order organizational learning. *Cybernetics and Human Knowing, 7(2-3)*, 47–69.

Steier, F., & Jorgenson, J. (2003). Ethics and aesthetics of observing frames. *Cybernetics and Human Knowing, 10*(3-4), 124–136.

Steier, F. (2005). Exercising frame flexibility. *Cybernetics and Human Knowing, 12*(1-2), 36–49.

Thomashow, M. (2020). *To know the world: A new vision for environmental learning*. Cambridge, MA: The MIT Press.

Watterson, B. (2005). *The complete Calvin and Hobbes: Book three*. Kansas City, MO: Andrews McMeel. (Originally published in 1995)

Lorusso, Mick. (2009). *Circle*. Urban Cosmos Series, Meditative Interventions.
Found object photograph. 25 x 30 cm.

Lorusso, Mick. (2016). *Red Mangrove*.
Photograph, part of Becoming Mangrove installation. 25 x 30 cm.

LoF22 – A Conference on Laws of Form
4-6 August 2022, University of Liverpool

Laws of Form is the title of the seminal work of George Spencer-Brown, first published in 1969. His book gave a new point of view on the role of distinction, markedness and the unmarked state in the creation of knowledge. This conference on Laws of Form is the second conference of its kind. The first conference was held in 2019, commemorating 50 years since G. Spencer-Brown's *Laws of Form* was published in 1969. A book corresponding to that first conference will soon be published by World Scientific.

Submissions for papers, panel sessions, interactive presentations, workshops, performance sessions, and creative contributions inspired by George Spencer-Brown's work and life—and particularly his key work, *Laws of Form* (LoF)—are now open and welcomed from participants keen to contribute to LoF22 which will be held from Thursday 4 August to Saturday 6 August, 2022 at the University of Liverpool. The url for information about the conference is **https://lof50.com/**

Keynotes will be given by Francis Jeffrey, Louis Kauffman, Barry Smith, and Stephen Wolfram. If you wish to actively participate at LoF22 please complete the submission form here: **https://easychair.org/my/conference?conf=lof22**

Facilities for giving presentations by video link will be available for those who, due to distance, are unable to attend in person. Presentations will be recorded and may be made available online. It is intended that, after the conference, as with LoF50, the papers will be compiled into a book.

Please email Florian Grote: **florian.grote@gmail.com** with any queries about the submission process.

Deadline for submissions: Monday 28 February, 2022. Late submission may be possible; contact Florian Grote.

There are no fees for attending the conference, for giving a presentation, or for submitting a synopsis of your paper.

Cybernetics and Human Knowing. Vol. 28 (2021), nos. 3-4, pp. 123–132

Virtual Logic—Recursive Distinctions

Louis H. Kauffman[1]

Introduction

This column is an introduction to recursive distinctioning (RD) (Isaacson, 1981; Kauffman & Isaacson, 2021). We first give a model and then discuss the cybernetics of describing describing in a context called *audioactivity* by John Horton Conway. We relate the simplest self-description

"Two Two's"

with both audioactivity and with ideas of self-reference related to the von-Neumann Machine that can build itself and the fixed points (eigenforms) of Church and Curry. We end with a reflective epilogue.

Recursion of distinctions creates a dialectic between meaning and syntax. By syntax I mean any formalism or language that has symbols, signs or distinctions. By meaning I mean to indicate our capacity for understanding understanding.

Meaning gives rise to syntax.
Syntax gives rise to meaning.

In the RD actions that we describe, meaning arises in the form of distinctions. These distinctions become signs, syntax for a next understanding in an endless round. It must be pointed out that mathematics is another kind of languaging, letting an observer play with symbols on paper without herself seeming to appear on that paper. In the course of that play, the observer is involved and not separated from the actions in the seemingly separate world of the paper. The meaning that this activity generates weaves a deeper world where the observer and her actions with paper or other media are inseparable from herself.

This paper is dedicated to the memory of Joel Isaacson and the memory of John Horton Conway and to their deep insight into fundamental process.

II. What is RD?

Recursive distinctioning means just what it says. A pattern of distinctions is given in a space or graph. Each distinction is seen as a sign and can be compared with its neighbors. Letters in a special alphabet are used to describe difference or sameness.

---

1. Email: loukau@gmail.com

Each sign is replaced by new signs that describe the structure of the distinctions made by the previous signs. The new signs describe the kind of distinctions among the old signs. The process continues recursively.

Here is an example. We use a finite row of letters. The special alphabet is

$$SA = \{ \ =, [,], O \ \}$$

where = means that the letters to the left and to the right are equal to the letter in the middle. Thus if we had AAA in the line then the middle A would be replaced by =. The symbol [means that the letter to the LEFT is different. Thus in ABB the middle letter would be replaced by [. The symbol] means that the letter to the right is different. And finally the symbol O means that the letters both to the left and to the right are different. SA is a tiny language of elementary letter-distinctions. Here is an example of this RD in operation where we use the proverbial three dots to indicate a long string of letters in the same pattern. For example, see Figure 1.

```
...  AAAAAAAAAABAAAAAAAAA  ...  is replaced by
...  =========]O[=========  ...  is replaced by
...  ========]OOO[========  ...  is replaced by
...  =======]O[=]O[=======  ...  .
```

Figure 1. The First Few Steps of Recursive Distinctioning.

Note that the element]O[appears from the simple difference between B and its neighbors, and that]O[then replicates itself in a kind of mitosis or DNA replication activity.

Recursive distinctioning is the study of those systems that use symbolic alphabetic language that can describe the neighborhood of a site (in a network) occupied by a given icon or letter or element of language. An icon representing the distinctions between the original icon and its neighbors is formed and replaces the original icon.

In Figure 2 we illustrate further steps in the recursive process (with a fixed boundary condition). Note the dialectical flavor of the continued patterning.

In this model, we have used synchronous processing so that each row is fully worked out before becoming the next row. It is convenient, particularly for pattern investigation, to use synchrony, but it is not necessary. Many asynchronous variations are possible, and we encourage the reader to explore these on her own.

RD processes encompass a very wide class of recursive processes. These elements are fundamental to cybernetics and cross the boundaries between what is traditionally called first and second order cybernetics. This is particularly the case when the observer of the RD system is taken to be a serious aspect of that system. Then the elementary and automatic distinctions within the system are integrated with the higher order discriminations of the observer. The very simplest RD processes have dialectical properties, exhibit counting and exhibit patterns of self-replication. Thus one has in the first RD a microcosm of cybernetics and perhaps, a microcosm of the world.

```
*AAAAAAAAAAAAAAAAAAAAABAAAAAAAAAAAAAAAAAAAA*
*                      ]O[                              *
*                     ]OOO[                             *
*                    ]O[ ]O[                            *
*                   ]OOOOOOO[                           *
*                  ]O[      ]O[                         *
*                 ]OOO[    ]OOO[                        *
*                ]O[ ]O[  ]O[ ]O[                       *
*               ]OOOOOOOOOOOOOOO[                       *
*              ]O[              ]O[                     *
*             ]OOO[            ]OOO[                    *
*            ]O[ ]O[          ]O[ ]O[                   *
*           ]OOOOOOO[        ]OOOOOOO[                  *
*          ]O[     ]O[      ]O[     ]O[                 *
*         ]OOO[   ]OOO[    ]OOO[   ]OOO[                *
*        ]O[ ]O[ ]O[ ]O[  ]O[ ]O[ ]O[ ]O[              *
*       ]OOOOOOOOOOOOOOOOOOOOOOOOOOOOOOO[              *
*     ]O[                              ]O[  *
```

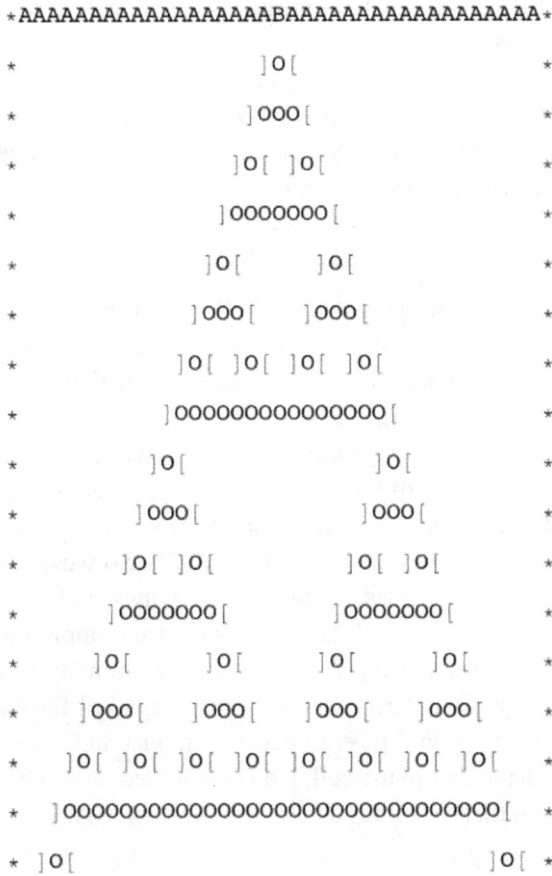

Figure 2. An extended RD Recursion with Boundary Conditions.

Back to the beginning of the RD and the analogy with DNA.
We have a sequence of letters such as
                    …BBBBBABBBB…
We will describe them in terms of their mutual likes and differences. Like this:

....*BBBBABBBB*...

...══⊃□⊂═══...

...══⊃□□□⊂══...

...═⊃□⊂═⊃□⊂═...

Figure 3. Focus on the basic RD.

If the letters on either side of a given letter are equal, I will replace the letter by an equals sign =. If the left side is equal but the right side is different, I will replace by ⊃. If the right side is equal, but the left side is different, I will replace by ⊂. If both sides are different, I will replace by a box □. Now we can perform recursive distinctioning. Examine the diagram above. We performed the distinction/description process three times, starting with …BBBBABBBB…. The change from B to A and back to B produced a *protocell* of the form

⊃□⊂

and then the next description elongates the cell, and in the next stage, the protocell divides into two copies of itself! All this comes from making distinctions and describing them with an alphabet so that one can make distinctions again and describe again. Very complex and interesting patterns can arise in this way.

This recursive distinctioning process then reminds us of DNA and how DNA replicates itself. You can think of the DNA molecule as a combination of two strands that we can call W (the Watson strand) and C (the Crick strand). W and C are chemically bonded and we can denote that by <W|C>. So we can write symbolically DNA = <W|C>. Special processes enabled by enzymes make it possible for these bonds to be broken and for the cellular environment to supply complementary base pairs to each separate strand. Letting E denote the environment we can write <W|E → <W|C> and E|C> → <W|C>. Here we have that the symbol for the environment can be erased or multiplied as in EE = E because it just indicates the possibility of interaction. Thus we have that in the cell, the DNA molecule can be separated into two strands, each of which then becomes a full copy of the DNA. In symbols this has the pattern:

$$<W|C> \rightarrow <W|E|C> \rightarrow <W|EE|C> \rightarrow <W|C> <W|C>.$$

Compare this symbolic sequence for DNA replication with the Recursive Distinction sequence we just discussed.

⊃□⊂→⊃□□□⊂→⊃□⊂=⊃□⊂

The interpretations are different, but the pattern is the same (to the generous eye). This is a place where RD needs research to reveal the deep structure indicated by this commonality of self-replication in RD process and DNA molecular process.

III. The Concept of RD

For us, the observers of the simple RD, there is an experience of recognition in seeing that this simple process mirrors the elementary processes of our own thought and discrimination. At that point of recognition the most fundamental problem arises: What is the source of the distinctions that we perceive?

On the one hand one can recognize that for a human observer a distinction is always accompanied by an awareness or consciousness of that distinction. Furthermore it is often the case that what is seen to be distinct depends upon the entire context of the event. A good example is the detection of the blind spot in the eye. This hole in our vision is normally not seen at all, but it can be revealed by looking in a direction to the left of a right thumb with the right eye (left eye closed). Then the thumb can disappear in the visual field, indicating the blind spot, but there is never a hole in the visual field. Some distinctions are distinctions for one modality of perception but not another. All distinctions that humans have are supported by their nervous system, biology and physics of the organism, and by the context in which these distinctions are framed. The context almost always involves a language of description, and that language itself is composed of distinctions.

The RD automaton is based in distinctions that arise in the contiguity of simple elements. In this case the elements are characters in symbol strings. The analogy can be carried forth to situations in cellular biology where the interactions are those of cells or constituents of cells, and the distinctions have to do with the direct interactions of molecules or with the making and breaking of cellular boundaries. In this arena we see that significant distinctions are seen to be in operation, apparently independent of our individual cognition and awareness.

This leads to the inevitable discussion of the notion of distinctions independent of human awareness. Such distinctions are understood by us to occur in other organisms and indeed within our own organism. So does the digestive system make its distinctions in regard to the food we hand it, and thereby enable the continuance of the body. So does my computer do its operations, independent of my possible understanding of its programming.

The RD automaton can suggest, in this field of analogies, that certain processes of distinction and indeed language precede the consciousness that we take to be the locus of distinctions for our understandings. Some reflection may convince the person who thinks about these ideas that the conception of distinction is circular. Distinctions beget distinctions in an endless round. And once again the RD automaton is a simple model of that dialectical process.

IV. The Audioactive Recursion

Examine the pattern of numbers below and, before reading further, find the rule that governs the pattern!

1

11

21

1211

111221
312211
13112221
1113213211

...

Illustrated here is a pattern of *recursive descriptions*. Each line is a description of the previous line. To see this, read the lines aloud. The second line says "one one" and that is a description of the first line. The third line says "two ones" and that is a description of the second line. The next line says "one two, one one", then "one one, one two, two ones" and so on. The full alphabet for this recursion is the set of numerals {1,2,3} and these are alternately signs and elements of the description of a pattern. This *audioactive sequence* was extensively investigated by John Horton Conway (1986) and has many mathematical properties.

A variant on the above recursion that is quite interesting is to start with the number 3 rather than 1. Then we have

3
13
1113
3113
132113
1113122113
311311222113

...

It is not hard to see that if the rows are $r_1, r_2, r_3, ...$ then r_{n+3} is an extension of r_n. The third row down from a given row is an extension of the given row. This means that we can build three infinite rows A,B,C that are in dialogue with each other in the sense that B describes A, C describes B and A describes C.

A = 111312211312...
B= 311311222113...
C= 132113213221...

There is much to explore in this recursion. A description is of course certainly a distinction, but the distinctions made by this form of description are of a more complex nature than the adjacencies in the first RD that we have discussed. Remarkably, the audioactive sequences shown here are based on a very small alphabet of numerals (1,2,3). It is a bit mysterious what can come from only one, two and three.

A = 11131221131211132221...

B = 31131122211311123113332...

C = 132113213221133112132123...

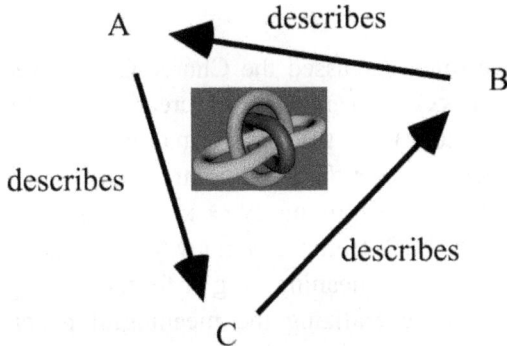

Figure 4 – A triplet description: B describes A describes C describes B.

We have two mappings defined on strings of digits that take strings of digits to strings of digits. For symbolic sake, lets let S denote the collection of all finite strings of digits. The D: S → S is our *descriptor* and U:S' → S is our *un-descriptor*. U is only defined on those strings S' that are descriptions.

Examine the sequence of digits 22. Its description is 22. Its un-description is 22. It is a perfect self-describer. D(22) = 22. U(22) = 22.

The description of Two Twos is Two Twos.
We can compare how 22 produces itself with
John von Neumann's machine B that can build itself!

The universal von Neumann machine B is a *universal builder*. Give B a description x and B will build the entity X with that description. So one would write

$$B, x \rightarrow X, x.$$

B would use the blueprint x to build X and produce X along with its blueprint x.

So it is fantastic. B can build itself. You just give B its own blueprint b! Then B,b → B,b and B produces a copy of itself.

$$B, b \rightarrow B, b.$$

Lets take the arrow nx → xxx...x (n x's) to mean the un-describe" arrow that produces the string whose description is nx. This is the analog of what a building machine does, and nx is the blueprint. Then we have 2x → xx and we see that 22 → 22 builds a copy of itself. This is of course a special case of von Neumann's pattern.

There is a 2 in the Von Neumann machine also. He has B, x → X,x. Two entities produce two entities. And the B,b is really a repetition just like 22 where the two 2's in 22 are different. One says the number of twos in the entity that is being described.

Fixed Points

In this column we have often discussed the Church-Curry fixed point construction where, given that gx = F(xx), then gg = F(gg), creating a fixed point or eigenform (Kauffman, 1987, 2005, 2012) for the transformation F. Here again we have the appearance of the number 2. In fact the fixed point construction is similar to defining nx → xxx...x (n x's) and then examining 2x → xx and discovering self-reference at the point of substitution 22 → 22. Here we begin with gx → xx and obtain gg→ gg by substitution, but there is no given meaning for g or for the arrow gx → xx.

It is very curious. By generalizing the meaningful pattern 2x → xx, to an apparently meaningless pattern gx → xx (from meaning to syntax) we enter in fact a wider domain of possible meanings (hence from syntax to wider meaning).

This is particularly so with the work of Church and Curry on the Lambda calculus since they generalize even further to the pattern gx = F(xx) and then obtain a fixed point for F by the substitution gg = F(gg). In fact, they were motivated in this endeavor by the special case of Rx = ~xx where ~ denotes negation in logic and AB denotes that "B is a member of A". Then Rx = ~xx is interpreted to mean that x is a member of R exactly when x is not a member of x. This is the Russell set. We obtain RR = ~RR, meaning that R is a member of R exactly when R is not a member of R. This is regarded as a paradox.

The form of this structure is that negation has a fixed point. In standard logic we do not allow fixed points for negation. But in other domains we certainly do allow forms of self-reference and fixed points. By pivoting across a pattern that can have many meanings (or none!) we obtain a wider view.

Cybernetics has the possibility of transcending individual disciplines by exploring the depths in the pure patterns of circularity.

Our own personal self-reference has a similarity to the 22. We realize that while we feel ourselves as me and myself (a two), this two is really a self-referring one. The unity of two and one arises from the self-reference of 22.

V. Epilogue

I should say a bit more about dealing with the line between what can be formalized, what is not yet formalized, and perhaps what cannot be formalized.

Distinction cannot be formalized. This is because a definition is an special form of a distinction. So any definition of distinction would be limiting the concept. This in no way inhibits us from pursuing distinctions. We have to understand that any given formalization is not everything. No model fully encompasses what it would purport to describe. No artifice will capture nature. No artificial intelligence will capture intelligence. And yet intelligent behavior can arise in the simplicity of recursive distinction.

Recursion arises when distinctions interact to produce new distinctions.

Distinctions arise and distinctions interact to produce distinctions.

Processes of this sort are at the base of all structure and the evolution of structure.

What are the fundamental distinctions?

Where do they come from?

Distinctions, both unaware and aware, arise. We pointed out that contiguous elements in strings or other grids may give rise to distinctions. In nature the action of such distinctions may have little to do with formalism.

Recursion in systems of distinctions tends to generate patterns of considerable complexity and relevance for what we observe in natural systems.

We have discussed in this essay the structure of recursive distinctioning and variants of it that are based on some or all of its themes. There is a need to find simple basic principles and constituents from which apparent phenomena can be built. Here it is proposed that distinctions are such elementals.

Distinctions (Spencer-Brown, 1969) escape the net of the conceptual. The conceptual is based upon fundamental distinctions. Distinctions escape the physical. For the physical connotes our on-going story of the distinction between feeling and form. No distinction in nature is independent of an observer. We see that the most elemental distinction, its indication and ourself are in the form identical.

The essential dialectic in the recursion of distinctions is a round where meaning begets syntax and syntax begets meaning. That circularity is the basis of the world.

Meaning → Syntax

Syntax → Meaning

This circularity generates meaning for second order cybernetics. Since we are at the end of his column, may this discussion be continued beyond the confines of these pages in the further actions of the author and his possible readers.

References

Conway, J. H. (1986). The weird and wonderful chemistry of audioactive decay. *Eureka, 46*, 5–16.

Isaacson, J. (1981). United States Patent 4286330, August 25, 1981.

Kauffman, L. H. (1987). Self-reference and recursive forms. *Journal of Social and Biological Structures, 10*, 53–72.

Kauffman, L. H. (2005). Eigenform. *Kybernetes, 34*, 129–150.

Kauffman, L. H. (2012). Categorical pairs and the indicative shift. *Applied Mathematics and Computation, 218*, 7989–8004.

Kauffman, L. H., & Isaacson, J. (2021). Recursive distinctioning and the basis of distinction. *Journal of Space Philosophy 10*(1), 69–82.
Spencer-Brown, G. (1969). *Laws of form*. London: G. Allen and Unwin.

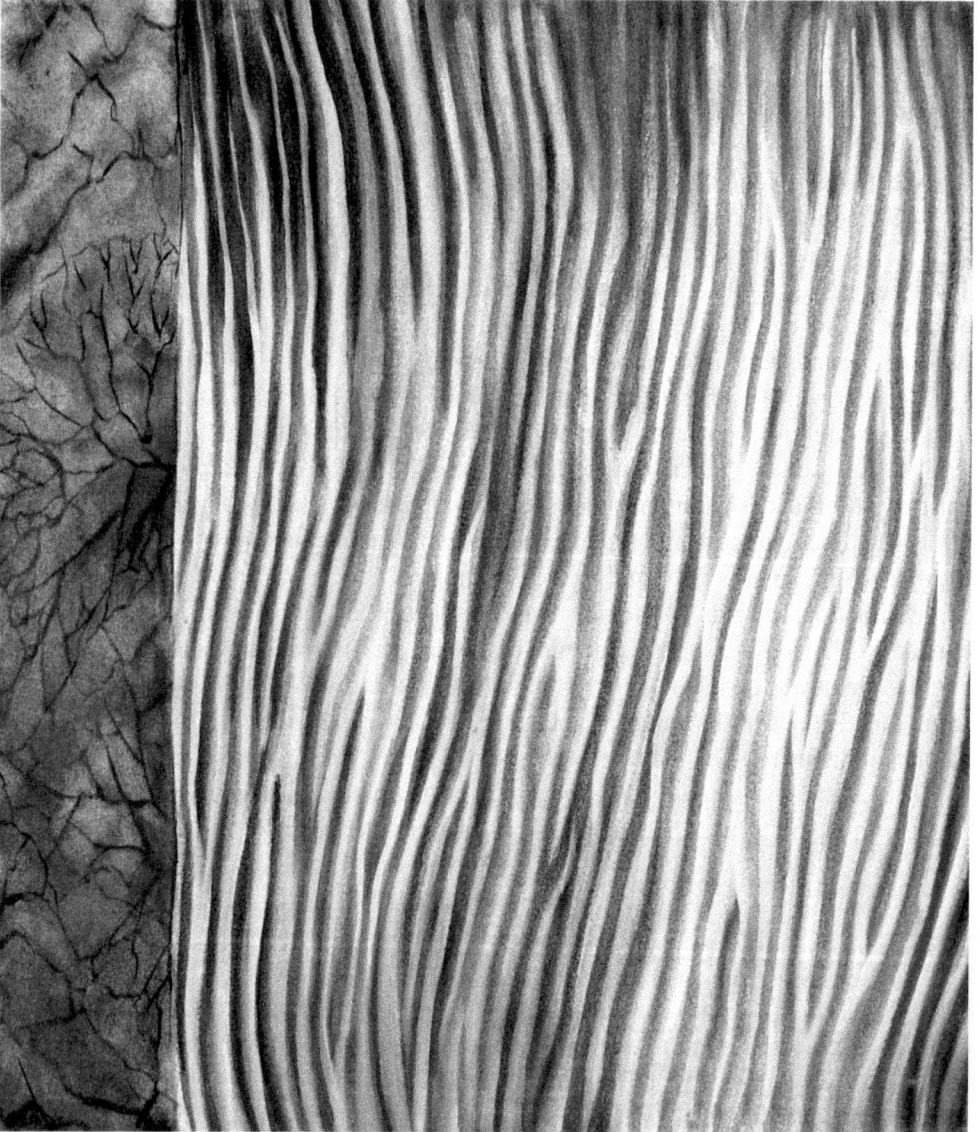

Lorusso, Mick. (2006). *Asariel*. Cosecha de Luz Series, Energy Patterns.
Painting. Oil on canvas. 75x80 cm.

Cybernetics and Human Knowing. Vol. 28 (2021), nos. 3-4, pp. 133–144

ASC

American Society for Cybernetics
a society for the art and
science of human understanding

Acting Cybernetically
in Complex Social Challenges
Designing for Sustainable Innovation

Goran Matic[1]

Introduction

Innovation participants are increasingly operating in the environments of social (DeLanda, 2019), political (Porte, 2015), organizational (Bertola & Teixeira, 2003) and ecological complexity (Ahern, Cilliers, & Niemelä, 2014). Attempting to devise strategies in such contexts challenges the "assumptions of order, of rational choice, and of intent" (Kurtz & Snowden, 2003, p. 462)—and presents an opportunity for designing sustainable engagements, that address needs of diverse stakeholders. Such engagements also need to support adaptations to *episodic* or *continuous change* (Poole & Ven, 2004)—in a way that enables viability of the anticipated innovation outcomes.

The feasibility of sustainable engagements may also be considered from the perspective of uncertainty—where, participants attempting to enact change often find themselves in *complex social challenge environments* (Matic, 2017). These contexts are characterized by the presence of *wicked problems* (Rittel & Webber, 1973), interactions between the *contextual, cognitive,* and *cooperative ambiguities* (Matic, 2017), and intrinsic asymmetries that manifest as emergent *paradoxes* (Andriopoulos & Lewis, 2010; Brown, Harris & Russell, 2010).

As such, complex environments necessitate innovation approaches capable of addressing imperfect knowledge (Head & Alford, 2015), cognitive blind-spots (Alrøe & Noe, 2014), and scenarios of information loss (Caron & Serrell, 2009). One possible explanatory approach is the Cynefin framework. Originally developed by Snowden (1999), it integrates concepts from the Complex Adaptive Systems (CAS) theory to posit that only certain classes of challenges can be thought of as problems that may have clear, heuristic or definitive responses.

1. University of Brighton. Email: G.Matic@brighton.ac.uk

In the Cynefin framework, simple challenges (such as cooking a meal) are seen to operate in the domain of the knowable, and can be addressed with prescriptive strategies akin to recipes. According to Snowden and Boone (2007) such contexts are "characterized by stability and cause-and-effect relationships that are easily discernable by everyone" (p. 2), and can leverage heuristics; as "often, the right answer is self-evident" (p. 2). Complicated issues (such as sending a rocket to the moon) operate in the environments where the unknowns may be reasonably anticipated. Snowden and Boone (2007) assert that such contexts "may contain multiple right answers, and though there is a clear relationship between cause and effect, not everyone can see it" (p. 3), necessitating approaches where one "must sense, analyze, and respond [which] often requires expertise" (p. 3). In contrast, complex issues (such as raising a child, innovating a product, or running a startup) and the chaotic ones (such as traversing an armed-conflict area) require very different strategies based on non-linear approaches. They require ways of adapting to unpredictable and uncertain environments.

Uncertainty and Disruptive Phenomena

In the wicked problem environments—where participants may not agree on what the fundamental issues are, or what they might be caused by (Rittel & Webber, 1973)—a possible strategy is to engage ecosystemic actors in some mode of *dialogic design* (Christakis & Brahms, 2003); or other methods of leveraging *collaborative rationality* (Innes & Booher, 2010).

A turn towards explicating the limitations of problem-solving approaches evolved from the work of early cybernetics and systems thinking theorists, made accessible by Horst Rittel and Melvin Webber, who taught, practiced and theorized on design, architecture and urban planning at the University of California at Berkeley.

Their influential 1973 paper, *Dilemmas in the General Theory of Planning*, posits wicked problems as a special category of challenges with ten distinguishing properties. The first two, "1. There is no definitive formulation of a wicked problem"; and "2. Wicked problems have no stopping rule" (Rittel & Webber, 1973, pp. 161–162) imply that it is impossible to define a wicked problem precisely, or decide when the work is completed, when attempting to implement change initiatives. The further criteria, "3. Solutions to wicked problems are not true-or-false, but good-or-bad" and "4. There is no immediate and no ultimate test of a solution to a wicked problem" (pp.162–163), posit active constraints around codifying or testing proposed solution approaches.

The further properties, namely "5. Every solution to a wicked problem is a 'one shot operation'; because there is no opportunity to learn by trial-and-error, every attempt counts significantly," "6. Wicked problems do not have an enumerable (or an exhaustively describable) set of potential solutions," and "7. Every wicked problem is essentially unique" (p. 164) assert a limitation to our understanding and the feasibility of implementing solutions, when it comes to applicability of previous experience.

Finally, the criteria "8. Every wicked problem can be considered as a symptom of another problem," "9. The existence of a discrepancy representing a wicked problem can be explained in numerous ways. The choice of explanation determines the nature of the problem's resolution," and "10. The planner has no right to be wrong" (p. 164), speak to the interconnectedness of complex phenomena, the arbitrary nature of explanatory power, and the escalated responsibility placed on designers.

To counteract these limitations, complexity-oriented innovation environments require involvement of "diverse and interdependent participants using authentic dialogue" (Innes & Booher, 2016, p. 8), as a mechanism of incorporating a richer dimension of views, perspectives and knowledge. Yet, multi-stakeholder environments are often fraught with issues resulting from clashes between the diverse perspectives, which can make community inclusion, stakeholder mobilization and participant engagement efforts challenging.

When the relationships between the key communities and multi-stakeholder groups become particularly tense or adverse, disruptive dynamics can emerge. One example is the phenomena of *schismogenesis*. It is a condition of progressive social deterioration described as a "process of differentiation in the norms of individual behavior resulting from cumulative interaction between individuals" (Bateson, 1958, p. 175). Derived from the Greek *skhisma* (split) and *genesis* (creation), the term *schismogenesis* describes the creation of a split between (or within) social groups—which causes progressively escalating disruption that can result in deep divisions. Yet Bateson observes that we must think of schismogenesis "not as a process which goes inevitably forward, but rather as a process of change which is in some cases either controlled or continually counteracted by inverse processes" (p. 190)

Cyclical Escalation and Social Coherence

Profound divisions in social groups are generally undesirable in community and innovation setting—as they tend to impact the potentialities of collaboration by destabilizing the sense of social coherence. Bateson (1958, p. 176) observes that "it is at once apparent that many systems of relationship, either between individuals or groups of individuals, contain a tendency towards progressive change." Bateson notes that if one social group engages in an assertive pattern while another is normatively expected to respond in a submissive manner, then it is "likely that this submission will encourage a further assertion, and that this assertion will demand still further submission" (p. 176). Such social dynamics can advance towards a state of affairs where "unless other factors are present to restrain the excesses of assertive and submissive behaviour" they might cyclically escalate towards a mutually reinforcing situation that affects "separate individuals or members of complementary groups" (p. 176).

Bateson applied these insights during World War II, to encourage desertion in the opposing forces. Messages were carefully crafted, then printed on pieces of paper and dropped from an airplane over contested territories. Their content was designed to polarize opinions and introduce doubts into the minds of soldiers—to create a sense of division and encourage attrition.

Whether used towards deliberate ends or occurring spontaneously, cyclical escalations can result in the emergence of disruptive *social drifts*—that may lead to breakdowns in the social fabric and disruptions in the sense of community coherence. In innovation contexts, this can affect the feasibility of implementing effective collaboration strategies—as per Fig. 1

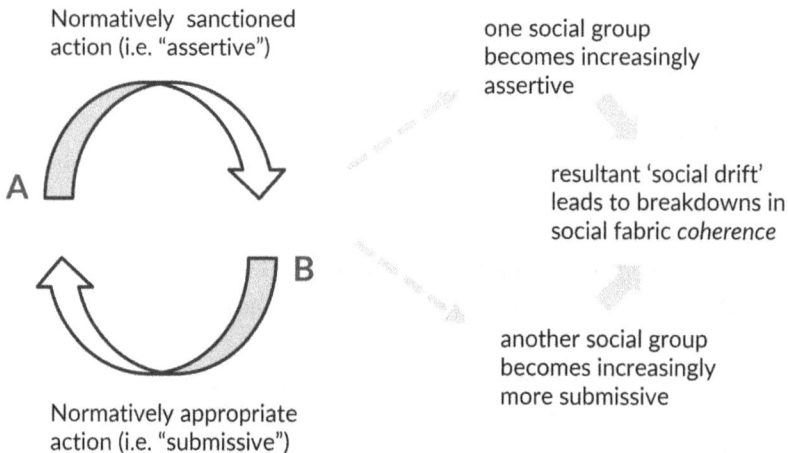

Normatively sanctioned
action (i.e. "assertive")

one social group
becomes increasingly
assertive

A

resultant 'social drift'
leads to breakdowns in
social fabric *coherence*

B

another social group
becomes increasingly
more submissive

Normatively appropriate
action (i.e. "submissive")

Fig. 1 – Cyclical Escalation: Social Groups Drift

Positive Feedback and Vicious Cycles

How do cyclical escalations work, and why might they be so impactful? They are a manifestation of the cybernetic concept of feedback loops, and an exemplification of positive feedback. This concept explains how rapid growth can occur through incremental amplification within a system—where each iteration builds on the previous one, in such a way where an output of an earlier iteration becomes an active input into the next one.

As such, positive feedback loops can rapidly grow small inputs towards large outcomes. When the deterioration of an underlying resource becomes fuel for accelerated cyclical growth—as in the case of forest fires (which burn hotter the more they burn) or the melting of the polar ice caps (which melt faster the more they warm-up)—positive feedback can turn into a vicious cycle, a situation where the deterioration of an underlying resource feeds into an overall acceleration of the loop.

When positive feedback turns into vicious cycles in a social context, the resulting cyclical escalations can create adverse effects in social coherence. This may lead to a range of anti-collaborative phenomena whose effects can extend beyond unequal social groups, and impact even the seemingly equal contenders. The associated emergent dynamics can manifest challenges in enacting sustainable innovation initiatives. One example is a dynamic pattern where overt activity from competitive actors emerges as continuously intensifying rivalry (Fig. 2).

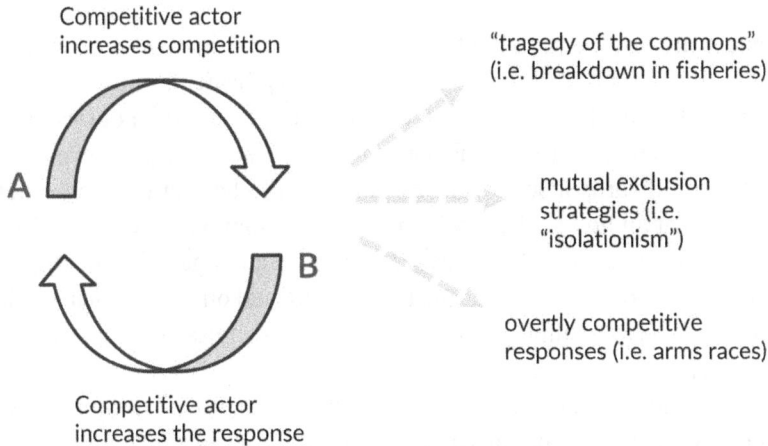

Competitive actor
increases competition

"tragedy of the commons"
(i.e. breakdown in fisheries)

A

mutual exclusion
strategies (i.e.
"isolationism")

B

overtly competitive
responses (i.e. arms races)

Competitive actor
increases the response

Fig. 2 – Cyclical Escalation: Increasing Rivalry

Social Coherence and Exergy

Social coherence disruptions impact the potential success and viability of innovation initiatives. However, it is often not clear as to how or why this might precisely be the case. To explore this further, it is useful to consider *exergy*—defined as the sum-total of available energy within a system which enables enactment of useful work (Wall & Gong, 2001). It is differentiated from the classical thermodynamic concept of energy in a sense that, while energy might in principle be present within a system, it might not actually be available for active use. From the research perspective, various phenomena can be interpreted as forms of social energy. For instance, according to Spiegelman, Spiegelman, and Spiegelman (2007) currency can be perceived as a tool which enables "pools of aggregated social exergy" (p. 270) and as a medium that "acts to skew information feedback loops between econosystem actors and larger scale structures such as economies and ecosystems" (p. 265).

This article posits that by creating conditions for the emergence of disruptive social phenomena, cyclical escalations degrade the available exergy—effectively reducing the total amount of accessible energy within a social system. Paradoxically, the potentiality of leveraging social energy might be there in principle—as manifested by such phenomena as the existence of goodwill, a shared desire to enact positive change, or even the availability of potential funding. And yet, such potentiality might be effectively inaccessible in practice, due to the degradations of the group sense of social coherence. This impacts the potentialities of being able to effectively engage in multi-stakeholder collaboration both across and within communities. In reducing the available social exergy, cyclical escalations also decrease the ability to collaborate and enact useful work in innovation initiatives.

Conversation as Design Strategy

To what extent might it be possible to design sustainable innovation initiatives if their environments are actively disrupted? This question is further complexified by the very nature of innovation challenges that can extend well beyond the notions of social drift and schisms of intensified rivalry. Environments where multiple stakeholders are in principle aligned towards a set of shared goals, while having different interpretative lenses around what an underlying problem might be, may not necessarily be engaging in direct rivalry; and might be experiencing challenges in identifying which interpretative perspectives are the most useful to act on. Examples might include governmental policy planning groups and Long-Term Disaster Recovery teams that operate in disruptive environments.

Disruptive phenomena often feature a wide range of *asymmetries* that are environmentally-oriented, and might be designated as scale, temporal, change or methodologically related (Matic, 2017). Asymmetries might also be thought of as stakeholder-oriented, and related to the expectation, attitude, cognitive, understanding or action differentials (Matic, 2017). The presence of multiple interacting asymmetries introduces fundamental ambiguities into the innovation environment, and therefore into the innovation process itself. One need not be deeply engaged in innovating within the domains of healthcare policy or social welfare to quickly arrive at the conclusion that the key issues in these areas are often intimately related to the notions around the availability of housing, feasibility of providing healthy food, accessibility of education and viability of gainful employment, for example.

The infeasibility of anticipating issues with certainty implies that effective innovation approaches must be able to dynamically adapt to the difficult-to-predict and potentially disruptive sets of conditions. This presents an opportunity to leverage the concept of social exergy. From this perspective, a key feature of an adaptive innovation process ought to be the capability of maintaining or enhancing the overall amount of social exergy throughout the engagements process. Such an approach supports social coherence, which enables the feasibility of collaborative engagements. And in doing so, it improves the likelihood of the overall innovation initiative being sustainable.

How can one enhance social exergy? A core inflection point is the stakeholder engagement process. Here, researchers observe that the "strength and nature of the link between stakeholder engagement, innovation management and entrepreneurship development is indisputably a critical question [where] stakeholder engagement for innovation management is a task of growing significance and the cornerstone of a win–win outcome" (Leonidou, Christofi, Vrontis, & Thrassou, 2020, p. 245). Links towards legitimizing innovation processes (Vershinina, Rodgers, Tarba, Khan, & Stokes, 2020) and achieving environmental sustainability to "solve complex problems, and gain social legitimacy" (Watson, Wilson, Smart, & Macdonald, 2018, p. 254) have been noted, while impacts on generating "knowledge for value cocreation" (Wiesmeth, 2020, p. 310) have also been established.

As such, stakeholder engagements need to be effective within and across the entire spectrum of complexity contexts. Yet they are also susceptible to an inherent design constraint. Due to the very presence of the pervasive ambiguity and uncertainty, the engagement processes cannot be deterministically designed in advance. To be viable, effective innovation strategies must generate options for ways of acting that are fundamentally capable of enhancing their own possibilities of success. This recursive definition signals an opportunity to leverage cybernetics methods to innovate innovation.

Existing approaches that support designing in complexity rely on specific theoretical underpinnings to infer opportunities for strategic action. Some of the more generative interpretative frameworks include wicked problems (Rittel & Webber, 1973), social messes (Ackoff, 1974), problematiques (Christakis, 2006) and the post-modern complexity (Cilliers, 1998). Irrespective of the interpretative lens, complex environments share a need to effectively involve a multiplicity of stakeholders. Thus, the innovation strategies for operating in complex environment typically leverage some mode of dialogic design (Christakis & Brahms, 2003) to effectively engage across the various communities of participation.

A shared reality in engaging complex environments, irrespective of the theoretical or interpretative frameworks used to understand them, is the notion of dialogue. Teams and practitioners need to use dialogue to converse with each other and with their participants and stakeholders. In this sense, dialogue can be viewed as instrumentalized, and as a primary medium of engagement within the environments of complexity. It can also additionally be perceived as a key enabler in inflecting the challenges identified within the innovation processes. Dialogue is capable of incorporating a diverse set of views, perspectives, and knowledge from interdependent participants (Innes & Booher, 2016) in an authentic manner. As such, it represents a key element in broader conversational structures that mandate specific design approaches and considerations. In that sense, dialogue can be considered as a form of conversation capable of enacting embedded re-construction, where the communication itself can be seen as a form of reflexive, embodied practice—capable of bringing about the sense of dynamic social coherence.

Conversation as Embedded Reconstruction
Can conversation be stated as having the necessary instrumental power to inflect the experience of social cohesion? Klaus Krippendorff has argued that conversation has recursive properties when posited as an embedded form of communication. Here, it is recognized that "social theories arise ... within a social fabric, constitutively involving human beings capable of inventing and articulating them" (Krippendorff, 1994, p. 79). From the recursive embeddedness perspective, conversation enables aggregative orders of meaning. This creates a capacity of creating reconstitutable networks—as

semi-persisted assemblages that can inform more complex and emergent structures, such as organizations. In this sense, the "reconstitutability of networks of conversation precedes all other criteria of the viability of organizational forms" (Krippendorff, 2008, p. 173).

This allows conversation to be posited as a possible candidate for the task of repairing disruptive social phenomena and for enhancing coherence. Here, conversation can be seen as a re-constitutive agent that is both interactive—in that it circumscribes a space "where cognitive phenomena surface: in the interactive use of language" (Krippendorff, 2009, p. 135)—and as manifesting a re-humanizing capacity; with a "recognition that speaking and writing are acts of continuously reconstructing reality" (p. 135). When conversation is viewed opportunistically—as an act of embedded re-construction—it can become a viable means of addressing disruptive social phenomena encountered in the environments of complexity. This enhances the feasibility of enacting sustainable innovation initiatives.

Communication as Reflexive Embodied Practice

In order to encompass social systems within the innovation process, conversation needs to have the necessary generative power to impact social coherence—as an emergent property of the aggregative set of relationships between social actors.

An additional characteristic enables this capacity, where conversation is recognized as both a medium and form of communication. From this perspective—and viewed cybernetically—conversation appears as reflexive in terms of containing potentialities of a "unified perspective that allows an observer to understand whole systems from their component parts and parts in the context of whole systems" (Krippendorff, 1985, p. 51). Krippendorff sees communication as embodied in a sense where "social theories … may also re-enter their social fabric and become embodied in the very practices of knowledgeable human agents" (Krippendorff, 1994, p. 79). Being recognized as both a reflexive and embodied practice, communication is posited as an intrinsic medium that informs the structuring of the social fabric, while intermediating the interactivity of the relationships within it. This positions conversation as an effective tool for inflecting the stakeholder engagement processes within the complex environments of uncertainty and ambiguity, pervasive in innovation landscapes. By inflecting conversation, we can also affect the continuous re-creation of the aggregative social structures, such as organizations—and in doing so, change them.

Conversation as Dynamic Coherence

How might dialogue and conversation be utilized as a tool to help stabilize social coherence in highly dynamic contexts? Additional perspectives can help us explore this question. Conversation theory (CT) conceptualizes the notion of coherence in terms of a process where conversation structures—that encapsulate key relationships in the context of shared understanding—are strengthened via local cycles that enable formation of dynamic *entailment meshes*.

An instance of an entailment mesh is a framework "in which potentially shared, commonly labeled public concepts can be related and within which participants can exhibit, for example, learning strategies, teaching strategies, analogy construction, or the use of analogy" (Pask, 1984, p. 17)—within which participants "must be able to add concepts and relations between them to the framework of concepts, and thus to make it evolve" (p. 17).

Here, coherence is viewed as a phenomenon where communicative elements become sufficiently stable to support more comprehensive forms and arrangements; the emergence of which enables dynamic coherence, a structural relationship which makes possible the creation of "iterative refinement of meaning" (Pask, 1992, p. 221). From this perspective, conversation can be viewed as a process of building dynamic coherence that continually engages stakeholders in the refinement of meaning.

While this theoretical perspective seems to support the notion of conversation as an effective means of stabilizing innovation processes, it also introduces another question. It implies a concern around discerning which entailment structures one ought to support when engaging in multi-stakeholder collaboration. This difficulty can be re-interpreted as a core challenge of establishing the necessary *directionality* of and within design. It could be re-stated simply as a need of continually re-directing conversation in the generative direction of enabling social coherence and successful outcomes of the associated innovation initiatives.

Design as Ethics

Questions of the directionality of design can be perceived as fundamentally ethical dilemmas. Here, the very notions around responsibility in the design process can be viewed through the lens of ethics as dialogue. Kenniff & Sweeting (2014, p. 1) suggest we "see ethics as consisting of a dialogue: the process of discovering, evaluating and contesting what is better or worse in any given situation, as opposed to the resolution of that particular situation in a way understood as being ethically 'correct.'" Such a process reframes design as "inescapably inter-subjective: a dialogue structured according to valued relationships between participants (both real and imagined)" (p. 2).

Viewing design as an implicitly ethical activity further reframes it as an emergent process where "it is not possible to fully analyse the situation in advance or to definitively frame the problem at hand because new questions, and with them new criteria, emerge in the process as the situation is explored" (Sweeting, 2015, p. 2). From this perspective, conversation and ethics can be viewed as instrumental methods of engagement where "the conversations that designers hold with other stakeholders … are not solely attempts to involve others for ethical reasons but part of how designers learn about the situation in which they act and the significance of what they propose" (p. 4).

Ethical conversation needs to be considered as a core element of enabling collaborative engagements and sustainable innovation in complex environments, as ethics can provide the necessary directionality in design processes. This can be accomplished with cybernetic ways of acting.

Acting Cybernetically

When manifesting as conversation structures, cybernetic / cyclic loops need not be viewed as capable of causing only disruptive social phenomena, such as schismogenesis. They can also be used for achieving reparative social and communication aims, while establishing a dynamic directionality within the design process itself. In doing so, the stakeholder engagements are designed from the recursive / self-reflective perspective by leveraging intrinsically ethical approaches.

Cybernetic ways of acting enable us to design generative potentialities for the emergence of cyclic structures capable of enhancing social coherence in complex settings. A key goal of such dynamic structures is to rehumanize social participation through the enacted inter-subjective processes. They can enable stakeholders to create entailment meshes capable of supporting sustainable innovation through conversation and dialogue. Participants are free to interact directly, validate truthfulness of others' claims, and rely on own experiences to form conclusions and construct shared knowledge through joint inquiry in a way that is interactive, reflexive, embedded and embodied. The associated cybernetic engagement model is presented in Figure 3, as per below:

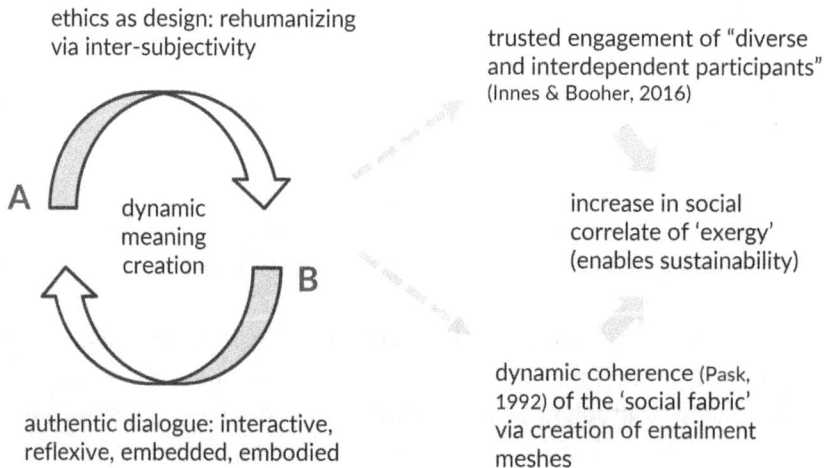

ethics as design: rehumanizing
via inter-subjectivity

trusted engagement of "diverse
and interdependent participants"
(Innes & Booher, 2016)

A   dynamic
meaning
creation   **B**

increase in social
correlate of 'exergy'
(enables sustainability)

dynamic coherence (Pask,
1992) of the 'social fabric'
via creation of entailment
meshes

authentic dialogue: interactive,
reflexive, embedded, embodied

Fig. 3 – Acting Cybernetically: Dynamic Meaning Creation

In this sense, multi-stakeholder engagements are posited to build trust between the diverse and interdependent participants (Innes & Booher, 2016) when collaboration is supported with continuous and adaptive re-interpretation of meaning through authentic dialogue. To achieve this, the sense of dynamic coherence (Pask, 1992) can enable a collaborative social fabric in a way that continually re-constitutes the associated entailment meshes in a generative manner, making communication possible.

The resultant enhancements of trust correspondingly increase social exergy. This further enables multi-stakeholder collaboration and adaptation within the environments of complexity and continuous change, thus mandating our efforts to act cybernetically.

References

Ahern, J., Cilliers, S., & Niemelä, J. (2014). The concept of ecosystem services in adaptive urban planning and design: A framework for supporting innovation. *Landscape and Urban Planning, 125*, 254–259. https://doi.org/10.1016/j.landurbplan.2014.01.020

Ackoff, R. L. (1974). *Redesigning the future: A systems approach to societal problems*. Wiley.

Alrøe, H. F., & Noe, E. (2014). Second-order science of interdisciplinary research: A polyocular framework for wicked problems. *Constructivist Foundations, 10*(1), 65–76.

Andriopoulos, C., & Lewis, M. W. (2010). Managing innovation paradoxes: Ambidexterity lessons from leading product design companies. *Long Range Planning, 43*(1), 104–122. https://doi.org/10.1016/j.lrp.2009.08.003

Bateson, G. (1935). 199. Culture contact and schismogenesis. *Man, 35*, 178–183. https://doi.org/10.2307/2789408

Bateson, G. (1958). *Naven: A survey of the problems suggested by a composite picture of the culture of a New Guinea tribe drawn from three points of view*. Stanford University Press.

Bertola, P., & Teixeira, J. C. (2003). Design as a knowledge agent: How design as a knowledge process is embedded into organizations to foster innovation. *Design Studies, 24*(2), 181–194. https://doi.org/10.1016/S0142-694X(02)00036-4

Brown, V. A., Harris, J. A., & Russell, J. Y. (2010). Collective inquiry and its wicked problems. In *Tackling wicked problems through the transdisciplinary imagination* (pp. 61–81). Earthscan.

Caron, R. M., & Serrell, N. (2009). Community ecology and capacity: Keys to progressing the environmental communication of wicked problems. *Applied Environmental Education & Communication, 8*(3–4), 195–203. https://doi.org/10.1080/15330150903269464

Christakis, A. N. (2006). A retrospective structural inquiry of the predicament of humankind. In J. P. van Gigch & J. McIntyre-Mills (Eds.), *Rescuing the Enlightenment from Itself: Critical and Systemic Implications for Democracy* (pp. 93–122). Springer US. https://doi.org/10.1007/0-387-27589-4_7

Christakis, A. N., & Brahms, S. (2003). Boundary-spanning dialogue for the 21st-century agoras. *Systems Research and Behavioral Science, 20*(4), 371–382. https://doi.org/10.1002/sres.508

Cilliers, P. (1998). Complexity and postmodernism: Understanding complex systems. Routledge.

DeLanda, M. (2019). *A new philosophy of society: Assemblage theory and social complexity*. Bloomsbury Publishing.

Head, B. W., & Alford, J. (2015). Wicked problems: Implications for public policy and management. *Administration & Society, 47*(6), 711–739. https://doi.org/10.1177/0095399713481601

Innes, J. E., & Booher, D. E. (2010). *Planning with complexity: An introduction to collaborative rationality for public policy*. Routledge. https://doi.org/10.4324/9780203864302

Innes, J. E., & Booher, D. E. (2016). Collaborative rationality as a strategy for working with wicked problems. *Landscape and Urban Planning, 154*, 8–10. https://doi.org/10.1016/j.landurbplan.2016.03.016

Kenniff, T.-B., & Sweeting, B. (2014). There is no alibi in designing: Responsibility and dialogue in the design process. *Opticon1826, 16*, Art. 1. doi: 10.5334/opt.

Krippendorff, K. (1985). Communication From a Cybernetic Perspective. *Informatologia Yugoslavica, 16*(1–2), 51–78.

Krippendorff, K. (1994). A recursive theory of communication. In D. Crowley & D. Mitchell (Eds.), *Communication theory today* (pp. 78–104). Cambridge, UK: Polity Press.

Krippendorff, K. (2008). Social organizations as reconstitutable networks of conversations. *Cybernetics & Human Knowing, 15*(3–4), 173–184.

Krippendorff, K. (2009). Conversation: Possibilities of its repair and descent into discourse and computation. *Constructivist Foundations, 4*(3), 135–147.

Leonidou, E., Christofi, M., Vrontis, D., & Thrassou, A. (2020). An integrative framework of stakeholder engagement for innovation management and entrepreneurship development. *Journal of Business Research, 119*, 245–258. https://doi.org/10.1016/j.jbusres.2018.11.054

Matic, G. (2017). *Collaboration for complexity—Team competencies for engaging complex social challenges*. Master Research Project (MRP), OCAD University. Retrieved April 30, 2019 from http://openresearch.ocadu.ca/id/eprint/1990/

Pask, G. (1984). Review of conversation theory and a protologic (or protolanguage), Lp. *Educational Communication and Technology Journal (ECTJ), 32*(1), 3–40. https://doi.org/10.1007/BF02768767

Pask, G. (1992). Correspondence, consensus, coherence and the rape of democracy. In G. van de Vijver (Ed.), *New perspectives on cybernetics: Self-organization, autonomy and connectionism* (pp. 221–232). Springer Netherlands. https://doi.org/10.1007/978-94-015-8062-5_13

Poole, M. S., & Ven, A. H. V. de. (2004). *Handbook of organizational change and innovation.* Oxford University Press.

Porte, T. R. L. (2015). *Organized social complexity: Challenge to politics and policy.* Princeton University Press.

Rittel, H. W. J., & Webber, M. M. (1973). Dilemmas in a general theory of planning. *Policy Sciences, 4*(2), 155–169. https://doi.org/10.1007/BF01405730

Snowden, D., & Boone, M. (2007). A leader's framework for decision making. *Harvard Business Review, 85,* 68–76, 149.

Sweeting, B. (2015). The implicit ethics of designing. In A. Ryan & P. Jones (Eds.), *Proceedings of the Relating Systems Thinking and Design (RSD4) 2015 Symposium.* Retrieved February 2, 2019, from http://systemic-design.net/rsd-symposia/rsd4-proceedings/

Vershinina, N., Rodgers, P., Tarba, S., Khan, Z., & Stokes, P. (2020). Gaining legitimacy through proactive stakeholder management: The experiences of high-tech women entrepreneurs in Russia. *Journal of Business Research, 119,* 111–121. https://doi.org/10.1016/j.jbusres.2018.12.063

Wall, G., & Gong, M. (2001). On exergy and sustainable development—Part 1: Conditions and concepts. *Exergy, An International Journal, 1*(3), 128–145. https://doi.org/10.1016/S1164-0235(01)00020-6

Watson, R., Wilson, H. N., Smart, P., & Macdonald, E. K. (2018). Harnessing difference: A capability-based framework for stakeholder engagement in environmental innovation. *Journal of Product Innovation Management, 35*(2), 254–279. https://doi.org/10.1111/jpim.12394

Wiesmeth, H. (2020). Stakeholder engagement for environmental innovations. *Journal of Business Research, 119,* 310–320. https://doi.org/10.1016/j.jbusres.2018.12.054

Lorusso, Mick. (2014). *Algae Capacitors 3.* Vodnik's Cells Series. Ecological Interactions.
Found materials photograph. 22 x 30 cm.

Cybernetics and Human Knowing. Vol. 28 (2021), nos. 3-4, pp. 145–152

The Face and the Machine

Devon Schiller[1]

A Review of Ksenia Fedorova's *Tactics of Interfacing: Encoding Affect in Art and Technology*
Cambridge, MA: The MIT Press, 2020. ISBN-13 9780262044158. 336 pp. $35 USD

Ksenia Fedorova's monograph, *Tactics of Interfacing: Encoding Affect in Art and Technology*, is a new entry in The MIT Press's Leonardo book series. It is an erudite, insightful, and sympathetic philosophical exploration of affective phenomena. Deeply thoughtful and well-paced, this book will call upon the reader to contemplate its contents long after they turn the last page. *Tactics of Interfacing* is thematically organized, with each chapter standing on its own and building upon one another, as Fedorova asks questions about intersubjectivity and selfhood through differing relations between affective phenomena and interface technologies. Topics addressed include: the face as a medium shared by both humans and machines; body image and the embodiment as well as algorithmizing of this image; the role that technologies play in transforming the relation between self and other; and, how the embodied self is situated within an environment. Collectively, these chapters offer an excellent, breathtaking exploration into the interface, which Fedorova importantly approaches "not so much as technology, but as a condition that brings to the fore and gives structure to the relational nature of being human" [p. 3].

A technopositivist meditation, Fedorova's *Tactics of Interfacing* belongs to what I would call the essayist tradition of media philosophy. That is, each of the individual chapters takes the form of an argumentative essay on a different but related topic, providing objective analyses as well as appraisal opinions. This anthology style is overall consistent with Anglo-American visual studies, as well as German-speaking *bildwissenschaft* (image science), and their immediate disciplinary progenitors (Rampley, 2012), as exemplified by the works of McLuhan, Mitchell, and Sontag, as well as Boehm, Kittler, and Warburg, among others. Also similar to such thinkers, Fedorova does not position her work "as an art historical investigation" [p. 5] in the traditional sense. Rather, *Tactics of Interfacing* is a philosophical study of cultural practices in which art serves "as a lens that brings together diverse philosophical and media conceptions of the aesthetic and ethical aspects of technologies' impacts" [p. 5]. Viewing her subject through the looking glass of Western phenomenological and political philosophy, Fedorova primarily focuses on Agamben, Arendt, Deleuze and Guattari, Heidegger, Lacan, Levinas, and Merleau-Ponty, among others. While at one and the same time, she also takes into consideration a variety of ideas from the behavioral and brain sciences, computer sciences, and semiotic sciences.

---

1. University of Vienna. Email: devonschiller@gmail.com

The kaleidoscopic diversity of thought in Fedorova's *Tactics of Interfacing,* a heterogeneity and a wandering that is, after all, her stated aim, constitutes one of the great strengths of the monograph. But it is also a constraint. On the one hand, Fedorova prioritizes theoretical inclusivity over theoretical framework. Consequently, *Tactics of Interfacing* would perhaps benefit from more consistent signposting before chapters and cross-referencing between concepts as well as from more considered historicity and contextuality behind the philosophies that are presented. All works of scholarship need not necessarily take the form of a so-and-so approach to this-or-that or a such-and-such investigation into these-or-those. Indeed, scientific endeavors would be dreadfully circumscribed and monotonous if they did. But, that Fedorova includes so many philosophers, yet so little on each of their bodies of work locates *Tactics of Interfacing* in the realm of generalist rather than specialist knowledge. On the other hand, Fedorova deploys this diversity of philosophy to good effect, with impressive breadth as well as with insightful depth. It is this that makes *Tactics of Interfacing* so highly relevant and significant. Whereas some media philosophers "appeal to art as a ground for developing their theoretical findings" [p. 5] Fedorova centers her argument on "modalities of experiences of relationality under the condition of their technological encoding" [p. 5]. In contrast to other recent works on the interactivity shared between organic and technological worlds, such as Katja Kwastek's *Aesthetics of Interaction in Digital Art* in 2013, or Inge Hinterwaldner's *The Systemic Image: A New Theory of Interactive Real-Time Simulations* published in German in 2010 and translated into English in 2017, both also from The MIT Press, Fedorova claims no novel theory. Rather, Fedorova acknowledges that "a model for the most balanced type of interfaciality … has yet to be found" [p. 36], and calls for the need to challenge the as-yet-unfinished principles of the interface. In such a way, Fedorova does the hardest work of philosophical inquiry in her *Tactics of Interfacing*, asking questions through direct address, questions that practitioners as well as theorists of art, science, and technology would do well to further reflect further upon.

One Chapter, for Example

For the purposes of this review, I will focus on the first of four chapters in *Tactics of Interfacing*—"Face to Interface"—as my privileged exemplar. In this chapter, Fedorova puts forward an aesthetics of equality in which humans and machines communicate with one another face-to-face, not necessarily as equals, but certainly through equivalencies. As Fedorova points out, the term "'interface' (at least in English) relates to 'face'; it is a zone in between two faces" [p. 15], with the word *face* etymologically derived from the "Latin *facia* (or *facies*) [which] means 'appearance' [and] also supported by the related Latin verb *facere*, 'to make,' 'to put something into form'" [p. 26]. From the perspective of an object-oriented ontology which very much aligns with the Heideggerian tradition, albeit the author herself does not articulate it thusly, Fedorova argues that both humans and machines have such a face, and while "the 'self' remains a category of the human" [p. 16], the machine is also an "entity with agency" [p. 16], which can be a subject or have subjectivity. "'Facing' one another [during human-computer interaction (HCI)], humans and machines need to quickly decode each other's respective actions and respond accordingly"[p. 23]. To facilitate this interaction, the "facial iconographies of digital systems" include not only face-like softwaric representations that invite "the expectation of an active response" [p. 23] but also "human-centered and anthropomorphic" [p. 23] modes of control. Fedorova begins her exploration of these modalities by tracing the paradigmatic changes to interfacial conditions in the digital age. Hand gestures controlled the graphical user interfaces of the late 1980s and 1990s, whereas in the human-face-to-machine-face model of the early twenty first century, as Anna Munster analyzes, the "need for the interface to act as a surface for translation is apparently lessoned if the human can be translated into a complex type of 'informational processing unit'" (Munster, 2006, p. 128; [quoted p. 24]). During such interfacing between anthropomorphized machine and mechanized human, Fedorova asserts, the human has to adopt the behavior and language of the machine, which itself has a preprogrammed emotionality that is neither "purely 'machinic'" [p. 24] nor actually complete with the complexities of emotional intelligence. Fedorova principally contemplates the expositional and mediational functionalities of anthropomorphic as well as machinic faces during certain interfacial conditions.

First, drawing upon the political philosophy of Hannah Arendt as well as Giorgio Agamben, Fedorova claims that faces "serve as a ground for commonality, as an indicator and a producer of the fundamental connectedness of human beings" [p. 26], and characterizes the face at least in part by the ways in which it constitutes a "language of diplomacy" [p. 27] that enables us to "*negotiate* the differences" [p. 27] within communities and environments. Arendt (1958/1998, pp. 198–199) describes the *polis*, a term which in Plato's *Republic* refers to a city-state based on the ideals of justice and virtue, as "the organization of the people as it arises out of acting and speaking together," and, although the face is not explicitly mentioned, a "space of appearance" in which people interact and which "precedes all formal constitution" of government and society. Similarly, but with an emphasis on the extralinguistic, Agamben (1995/2000, pp. 90–91) describes how the "face is the only location of community, the only possibly city," because what the face "exposes and reveals is not *something* that could be formulated as a signifying proposition of sorts," but rather the "face's revelation is revelation of language itself." Whether understood as Arendt's *appearance* or Agamben's *exposition*, Fedorova suggests that such a public face "allows one *to take possession of the exposition* of oneself, and more precisely—possession of not only *what* is being exposed, but the fact of the exposition" [p. 27], a process which has perhaps never been so intrinsic and influential as in this era of big data surveillance and social media spectacle.

Then, drawing upon the phenomenology of Emmanuel Levinas as well as Gilles Deleuze and Félix Guattari, Fedorova challenges "the (Platonic) representational model of a face" [p. 28]. As Fedorova claims, "what matters is not only *what* is being communicated, but the act of opening, exposing one's self (the image of oneself) to the other as an invitation to go beyond such distinctions (self and other), entering the space of exchange" [p. 28], that is, of interfacing. Levinas (1998, pp. 50–51) describes how the "face of the Other at each moment destroys and overflows the plastic image it leaves me." And Deleuze and Guattari (1987) describe the ways in which human faces as well as nonhuman objects are "engendered by an *abstract machine of faciality*" (p. 168), which "carries out the prior gridding that makes it possible for the signifying elements to become discernible" (p. 180), but which "does not assume a preexistent subject or signifier" (p. 180). Throughout her explication of the interfacial, Fedorova employs terminology from semiotics, such as when she references what she calls the "classical sender-message-receiver model" [p. 28; see also pp. 198–199] from the structuralist tradition of Saussure and Jakobson. Yet as happens all-too-frequently in media philosophical writing today, Fedorova seems to rely on the watered-down semiotic theory that subsists in communication studies and media studies. Not only does Fedorova simplify this model, its complexity and variations, and the factors of communication therein addressed, but she makes no mention of the extensive literature in computer semiotics and cybersemiotics on co-operative, embedded, and interfacial sign systems.[2] Despite this oversight, Fedorova nimbly avoids reducing the face, and its numerous communicative functionalities, to a type of media. Rather, Fedorova explores the "sense of selfhood and identity that the face conveys" [p. 36], how the "face serves as a tool for making distinctions between self and other" [p. 36], and, using terms from Lacan not Uexküll, the ways in which the face positions subjects within their "*Innenwelt* (organic inner world) and *Umwelt* (external reality)" [p. 47] on both intra- and interindividual levels.

Having established her theoretical grounds, Fedorova curates her case studies into resemblance clusters rather than rigid classifications. These case studies include media artworks from the genres of installation art and interaction art which are based on the electrical stimulation of the human face, the compositing or superimposition of facial images, and simulated exchanges that stimulate empathy and introspection. For instance, Fedorova explicates the ways in which Megan Daalder's *The Mirrorbox* in 2010, Mattia Casalegno's *Unstable Empathy* in 2012, and Karen Lancel and Hermen Maat's *Saving Face* from 2012 to 2017 "make the discrepancy between the singular and the generic felt through the physical experience of having an image of one's face superimposed on the image of another person's face" [p. 26]. This contrasts with the composite portraiture of eugenicist physiognomics, such as those by Francis Galton in the late nineteenth century, a composite which has a "tautological relation with its mental origin," Josh Ellenbogen (2012, p. 122) points out, and should in and of itself "tell us of its typicality [and] if the images that go into it really belong to the same class" (p. 122). Fedorova suggests that "the technique of superimposition, or morphing participant's faces into one another's with the resulting effect of merged identities" [p. 36], is applied in these artworks in order to analyze "issues of fluid and

---

2. For example: Clarisse Sieckenius de Souza, *The Semiotic Engineering of Human-Computer Interaction*; Frieder Nake, "Surface, Interface, Subface: Three Cases of Interaction in One Concept"; and, Shaleph O'Neill, *Interactive Media: The Semiotics of Embodied Interaction* . See also: Charles Goodwin, *Co-Operative Action*.

networked identity, transference of trust and affiliated feelings of empathy and vulnerability" [p. 36]. In each such artwork, Fedorova shows, a *hybrid face* emerges, and consequently a "sense of a hybrid identity" [p. 39], for which the interface is "the place where bonding is experienced" [p. 39] and potentials are realized. Across this chapter, Fedorova mindfully compares these contemporary artworks to their historical predecessors and successors, also with Arthur Elsenaar's facial choreography and Guillaume Duchenne's physiognomic orthography, as with Alexa Wright's *Alter Ego* and Affectiva's emotionAI, among others. But what is perhaps most refreshing in Fedorova's *Tactics of Interfacing* is the absence of the apocalyptic in her take on technological acceleration, and her assessment of humanity not as bungled and botched, as Nietzsche wrote, but as active and agential. As Fedorova concludes, "while enmeshed in perpetual negotiations between machinic codes and one's own nature, a human still remains free to imagine and continuously reimagine oneself and the surrounding world" [p. 246].

Further Discussion and Future Research

In addition, *Tactics of Interfacing* brings to light several methodological problems in media philosophy and, particularly, in media philosophy that is based to one degree or another on media art. These problems do not in any way take away from the superior quality of Fedorova's monograph. But I suggest that they should be considered as we proceed together as a community into the horizon of the field.

Firstly, Fedorova supports her argument with findings from the behavioral and brain sciences of psychology and neurobiology. These findings involve the default-mode network and mirror neuron system [see pp. 43–45], which help to facilitate the understanding of emotions, goals, and intentions in others and regulate interaction between individuals, as well as the ELIZA effect [see pp. 144–145], the nonconscious tendency to anthropomorphize machines and analogize machinic to human behaviors. Certainly, the concepts of the DMN, MNS, and ELIZA effect are highly relevant to the context of *Tactics of Interfacing* and are cited appropriately and comprehended adeptly by Fedorova. Yet, this material pops up in the text occasionally rather than essentially, more isolated sidebar than integrated evidence. And there is always a risk that any such appropriation from the sciences to the humanities be superficial at best or fallacious at worst. The treatment of behavioral and brain science in this work of media philosophy ascends beyond the elevation of the Wikipedia-esque, as is noticeable in the notes, for example.[3] However, the potential for this progress is impeded by the application of an *argumentum ab auctoritate* (argument from authority), in which Fedorova appeals to the authoritative sources of this dominant paradigm, such as Damasio, Gallese, Lakoff, and Rizolatti, to justify her interpretation of the psychological and physiological effects of interactive installations.

---

3. See Fedorova [p. 254 n47 & p. 276 n13].

Do not misunderstand me, I am never more satiated than when an author serves neuroscience with their philosophy. And as a semiotician in the humanities, I myself share a proclivity for sprinkling rather than stirring neuroscience into the mix. But as the media philosophy of media art continues to shift from the alternative to the mainstream and from the new to the norm, I cannot help but hunger for multilateral interdisciplinarity over disciplinary Frankensteining, the true over the topical. Obviously, each and every work of scholarship must hierarchize and prioritize its material. And in Fedorova's *Tactics of Interfacing*, behavioral and brain science takes second or even third place to phenomenological and political philosophy. Not to romanticize, but in the media philosophy of the mid-to-late twentieth century, scholars not only called upon but also contributed to the field of psychology, as with Mulvey's male gaze or Sontag's voyeurism of suffering. And today, there are thinkers writing on both sides of the two cultures divide who equally develop and fully deploy the potential links between art and brain, from the neuroarthistory of art historian John Onians, to the aesthetic reductionism of neuroscientist Eric R. Kandel. Can media philosophy facilitate productive exchange between the brain science and cultural studies, and if so, how?

Secondly, Fedorova relies extensively on artist statements in her critical analysis of the conditions of interfaciality, such as with Elsenaar's *Face Shift*, Daalder's *The Mirrorbox*, and Gonsalves' *Chameleon*,[4] including both integrated and block quotations. Fedorova deeply reads these texts, challenging but not contradicting the assertions made by the artists, and taking them as the point of departure through which she begins her exploration. However, artist statements pose a number of challenges for the historian as well as the theoretician because of their self-representativeness. Like other kinds of first-person ego documents, such as diaries or letters, artist statements are written for a particular audience. Consequently, artist statements can potentially be coded with anecdotal, community-specific, or personal sign systems. Down through history from Da Vinci's notebooks to Van Gogh's letters, artists have used the written word, in all its digitalness and explicitness, to clarify and to contemplate their art. But of course, an artwork does not necessarily do what an artist wants it to or says it does. An artwork's meaning is not limited to the artist's intention. And, ultimately, each and every interpretation resides in the eye of the beholder.

In *Tactics of Interfacing*, Fedorova demonstrates the ways in which media art takes the traditional practice of the artist statement to a whole other level, that of a hybrid genre which blends the artist statement with scientific paper and technical documentation. As Christiane Paul (2016) specifies, media art is sometimes made by a single artist or multiple co-artists in a studio, but is more frequently created by an interdisciplinary collaboration, one which involves creative directors, software developers, technical designers, and other specialists, and originates in an institute, laboratory, or university. Given this foundational character of media art, the paramedia that accompanies the artwork and informs its interpretation may include, for instance,

---

4. See Fedorova's discussion of Elsenaar [pp. 32–35], Daalder [pp. 37–38], and Gonalves [pp. 58–64].

the artist statement and other first-person textual descriptions of the work, as well as explanatory diagrams, installation models, technical schemata, and so on. Further, this documentation can often be published as a scientific paper, which presents experimental results with technical language in a peer-reviewed journal, such as *Leonardo*, also from The MIT Press. Yet, cross-inter-multi-transdisciplinarity is a spectrum not a state. And some media artists know more about the science and technology that they do more than others, considering the number of Sunday painters in the media arts who make use of the equipment of their workplace, as well as the increasing proliferation of black box technology solutions. Reliance on artist statements can reify artists, and in turn ahistorically revise their life and works into those of the archetypal genius melancholic touched by the daemon of inspiration, as Vasari codified, Panofsky critiqued, and today has been nigh-mythically personified in the great masters of the Renaissance and the cults of personality of the 1960s and 1970s. This is not to say that creativity, inspiration, and talent do not play a part in making art, or to reduce the aesthetic to the hormonal or the synaptic; every "ah-ha!" moment, "eureka!" cry, and "f#@k!" failure-turned-success has its place. But if the artist statement in and of itself is not reflected upon critically and critiqued with reflexivity, then the robustness of the philosophizing about what is actually within the picture plane of the artwork, however embedded or extended, will be the worse for it. How does a media art historian, philosopher, or theorist evaluate the evidence in an artist statement as a historical document and primary source, which is context dependent, and which is not value neutral?

References

Agamben, G. (2000). The face. In *Means without end: Notes on politics* (V. Binetti & C. Casarino Trans., pp. 91–102). Minneapolis, MN: University of Minnesota Press. (Originally published in 1995 as *Mezzi sensa fine* by Bollati Boringhieri editore s.r.l.)

Arendt, H. (1998). *The human condition* (2nd ed.). Chicago: University of Chicago Press. (Originally published in 1958.)

de Souza, C. S. (2005). *The semiotic engineering of human-computer interaction.* Cambridge, MA: The MIT Press, 2005.

Deleuze, G., & Guattari, F. (1987). *A thousand plateaus: Capitalism and schizophrenia* (B. Massumi, Trans.). Minneapolis, MN: University of Minnesota Press.

Ellenbogen, J. (2012). *Reasoned and unreasoned images: The photography of Bertillon, Galton, and Marey.* University Park, PA: Pennsylvania State University Press.

Favareau, D. (Ed.). (2018). *Co-operative engagements in intertwined semiosis: Essays in honour of Charles Goodwin.* Tartu, Estonia: University of Tartu Press.

Fedorova, K. (2020). *Tactics of interfacing: Encoding affect in art and technology.* Cambridge, MA: The MIT Press.

Goodwin, C. (2017). *Co-operative action.* New York: Cambridge University Press, 2017.

Hinterwaldner, I. (2017). *The systemic image: A new theory of interactive real-time simulations* (E. Tucker, Trans.). Cambridge, MA: The MIT Press. (Originally published in 2010 as *Systemische Bild* by Wilhelm Fink Verlag.)

Immanuel L. (1998). *Totality and infinity: An essay on exteriority.* (A. Lingis, Trans.). Pittsburgh, PA: Duquesne University Press, 1998.

Kwastek, K. (2013). *Aesthetics of interaction in digital art.* Cambridge, MA: The MIT Press.

Munster, A. (2006). *Materializing new media: Embodiment in information aesthetics.* Hanover, NH: University Press of New England.

Nake, F. (2008). Surface, interface, subface: Three cases of interaction in one concept. In U. Seifert, J. Hyun Kim, & A. Moore (Eds), *Paradoxes of interactivity: Perspectives for media theory, human-computer interaction, and artistic investigations* (pp. 92-109). Bielefeld, Germany: transcript Verlag.

O'Neill, S. (2008). *Interactive media: The semiotics of embodied interaction.* London: Springer-Verlag.

Paul, C. (2016). From digital to post-digital: Evolutions of an art form. In C. Paul (Ed.), *A companion to digital art* (pp. 1–20). Chichester, UK: Wiley Blackwell.

Rampley, M. (2012). Bildwissenschaft: Theories of the image in German-language scholarship. In M. Rampley, T. Lenain, H. Locher, A. Pinotti, C. Schoell-Glass, & K. Zijlmans (Eds.), *Art history and visual studies in Europe: Transnational discourses and national frameworks* (pp. 119–134). Leiden, Netherlands: Koninklijke Brill.

Lorusso, Mick. (2006-08). *Bursting Forth*. Anima Mundi Series, Energy Patterns.
Drawing. Graphite on paper. 43 x 35 cm.

www.ingramcontent.com/pod-product-compliance
Lightning Source LLC
Chambersburg PA
CBHW080425270326
41929CB00018B/3161